Praise for Alan Connor

'Alan knows everything, knows everyone,
and writes beautifully too.'
Richard Osman

'I've always been in awe of Alan Connor – crossword maven, quintessential quizzer and the man with the contents of the *Oxford English Dictionary* stored just above his left eyebrow . . . and he's quite funny too.'
Rory Cellan-Jones

'A gorgeous, funny tour of the British Isles as seen from the clouds. Definitely a book for all weathers – including rain. Obviously.'
Konnie Huq

'[An] absolute delight'
Mail on Sunday

'charming, fascinating'
Sunday Times

'amusing and informative'
Washington Post

Also by Alan Connor

The Shipping Forecast Puzzle Book
Pointless Facts for Curious Minds
Richard Osman's House of Games
House of Games: Question Smash

188 WORDS FOR RAIN

A delightfully damp tour of the
British Isles, led by natural forces

ALAN CONNOR

BBC
BOOKS

BBC Books, an imprint of Ebury Publishing
Penguin Random House UK
One Embassy Gardens, 8 Viaduct Gardens,
Nine Elms, London SW11 7BW

BBC Books is part of the Penguin Random House group of companies whose addresses can be found at global.penguinrandomhouse.com

Penguin Random House UK

Copyright © Alan Connor 2024

Illustrations © David Wardle

Alan Connor has asserted his right to be identified as the author of this Work in accordance with the Copyright, Designs and Patents Act 1988

Penguin Random House values and supports copyright. Copyright fuels creativity, encourages diverse voices, promotes freedom of expression and supports a vibrant culture. Thank you for purchasing an authorized edition of this book and for respecting intellectual property laws by not reproducing, scanning or distributing any part of it by any means without permission. You are supporting authors and enabling Penguin Random House to continue to publish books for everyone. No part of this book may be used or reproduced in any manner for the purpose of training artificial intelligence technologies or systems. In accordance with Article 4(3) of the DSM Directive 2019/790, Penguin Random House expressly reserves this work from the text and data mining exception.

First published by BBC Books in 2024

www.penguin.co.uk

A CIP catalogue record for this book is available from the British Library

ISBN 9781785948541

Typeset in 11.75/19.7pt Adobe Caslon Pro by Jouve (UK), Milton Keynes
Printed and bound in Great Britain by Clays Ltd, Elcograf S.p.A.

The authorised representative in the EEA is Penguin Random House Ireland, Morrison Chambers, 32 Nassau Street, Dublin D02 YH68

Penguin Random House is committed to a sustainable future for our business, our readers and our planet. This book is made from Forest Stewardship Council® certified paper.

Contents

1	A Mizzle Comes to Dartmoor	1
2	Raining Knives and Forks in Snowdonia	19
3	A Fliuch Mourne Morning	37
4	Just Spittin' (A Manchester Downpour)	55
5	The Solway Firth Disappears in Scotch Mist	73
6	Spairging in the Celtic Rainforest	93
7	'We're Sure to Have a Plash' in Northumbria	111
8	Plothering Again in the Midlands	131
9	There's a Bange in the Fens	151
10	And in Kent, More Folkestone Girls than Ever	171

Acknowledgements	193
List of Rain Words	195
About the Author	201

1
A MIZZLE COMES TO DARTMOOR

You're in the sky, seven thousand feet above the bottom bit of the British Isles. You gaze down first at the figures lining and crossing the heaths and bogs of Dartmoor. They gaze up.

'Looks like rain.'

They're right, of course. They will spend, we're told, five months over the course of their lives discussing whether it's going to rain. They might or might not have forgotten the clouds' Latin names; they might or might not have ever properly understood how they appear above them and why they sometimes suddenly disappear.

But they can tell from a glance which of the masses of water up here will stay up here, and which will come down and make them wet. Without thinking, in the same moment, they pick up on dozens more cues and clues from the air, from plants and animals and from everyday things in their kitchens – clues we will return to.

And the rain will brighten their day.

There on the edge of the moor, a picnic will take a different form. The decision will be made by the **mizzle**. The picnickers are comers-in, not from these parts and here on holiday, and they don't call it mizzle. But mizzle is the local word in sundry spots and, right now, it's the local weather. It's also our **rain word #1**.

In 1983, weatherman Francis Wilson joined the *BBC Breakfast* team and gave forecasts in a new way. The graphics were generated by computers, replacing the magnetic clouds that viewers were accustomed to seeing. (They were also accustomed to hoping that today might be one of the days the clouds would slide comically off the map.)

Francis, as the nation knew him, also brought with him a less formal set of words to describe the weather. Not for Francis the *th*under*storm* when he could instead tell of a coming 'thorm'.

Likewise, mist that's apt to turn to drizzle was, in Francis's reports, some mizzle. But this term was much older than Francis. In Dutch, the word for this kind of rain has long been 'misel'. And 'mizzle' is older even than 'drizzle'. So old, we find it in one of the first books in English: Caxton's translation of Virgil's *Aeneid*. An already-challenging journey is made more difficult by 'mysell', followed by 'grete heyle stones'.

In the 1490s, mizzle was a more aggressive rain than the kinds that are nowadays called mizzle in the West Country. Mizzle has been applied to milder and milder rain over time. By the nineteenth century, we find it handily placed in a sequence of types of rain in a wildly popular book called *The Miseries of Human Life*, a catalogue of 'petty outrages, minor humiliations, and tiny discomforts that make up everyday human existence'.

One petty outrage described is the experience of setting off for dinner without the right coat . . .

> . . . *in a mist, which successively becomes a mizzle, a drizzle, a shower, a rain, a torrent.*

Like that sodden Georgian dinner guest, our would-be picnickers, waiting in their car, once again consider that adage: there's no such thing as bad weather, only the wrong clothing.

> — Who was it that said that? The Lake District guy, who wrote the books?

— I thought it came from an ad for Gore-Tex?

— Don't be obtuse. What was his name? With the handwriting I said you could learn a lot from? Look it up, would you?

— I can't look it up, remember? If we had a signal, we'd have been able to check the forecast.

— I *did* say we should have checked the forecast before we left...

Then it's quiet in the car for some long moments.

The light dims just a little as the mizzle rapidly becomes **#2: drisk** (a misty drizzle that makes everything seem to shimmer) and then **#3:** an actual **drizzle**. In the most technical of terms, a drizzle comes in droplets of a diameter of less than half a millimetre, but if it's moving too fast for you to measure with a ruler, it's also rain that won't splash in puddles. In practical terms, the decision is confirmed. It's going to be a car picnic.

As the cool bag is unzipped, as scotch eggs are passed around and as breadcrumbs begin to decorate the upholstery, they quickly remember: they enjoy a car picnic. Often, they prefer it. They recall the car picnics of the past: always with the occasional beat of the windshield wipers to offer a view of the drisky landscape; always with a vow to remember to later use the hand-held vacuum cleaner to remove the breadcrumbs; always with a reminder that the hand-held vacuum cleaner can handle nothing more substantial than dust. And then the Lake District name is recalled: Wainwright.

Steamy, punctuated by pitter-patter, car picnics really are the best. The picnickers are lucky to be given one: unknown to them, the next road along is rainless. If they were there, they'd have missed their chance, stuck in the dry and windy quiet. So: where does this rain stop? Unlike with some, it's impossible to say. This rain, this mizzle-become-drizzle, is too fine and weak to have a proper edge. You couldn't find an edge, even from your vantage point above the contented picnickers.

A vague quality not possessed by the kelsher up on the moor.

Like many of our rain words, **kelsher (#4)** is not said only in the West Country. It's used in Lancashire, it's used by earth scientists and mineralogists, and it always means the same thing.

For rain to be a regular kelsher, it needs to be heavy and brought by a strong wind. It's one of those pieces of language that tells you by its sound what it's up to. If you'd never heard the word and someone told you to expect them to come home having been kelshed, you wouldn't expect a mild spattering on their clothing.

In 1955, the kelsher's kelsher – the heaviest kelsher of all – fell the other side of the Blackdown Hills and the problem wasn't the wind that brought it: it was the lack of wind once it arrived. In summer 1955, the Dorset village of Winterborne St Martin saw temperatures close to 30°C for days on end. Gardeners yearned for rain, but you can have too much of a good thing.

At the same time, heading for the Channel was a Spanish plume: a very warm air mass from the Iberian peninsula which, when it reaches our part of the world, goes up, as warm air is wont to do. In this instance, it formed a lid over a package of incredibly unstable air.

And so the Spanish plume brought to Winterborne St Martin cumulonimbus clouds and all that comes with them: thunder, lightning and lots and lots of rain. After it arrived, the wind ominously and suddenly dropped – and the plume stayed over Winterborne for three days. In one 15-hour period, 11 inches fell. Roads became rivers, a dam burst, and all the rest of it. While a

A MIZZLE COMES TO DARTMOOR

Spanish flume is often a welcome arrival, the only decent thing that can be said for being stuck under one for days is that you should end up with better defences for any future floods, though there has been none yet to match 1955's kelsher.

For the pair of walkers you now see below you, today's kelsher is no trouble. Most importantly, there's plenty of wind today, so the rain will be shared across the moor. They too have no idea who first said that there's no such thing as bad weather, only the wrong clothing, but they've said it themselves and taken heed, so, apart from their faces, they're warm and completely dry.

Most happily for them, the kelsher has deterred other would-be walkers. They have the place to themselves.

Insects emerge as rain encourages them to explore the floor of the moor, a development that doesn't escape the attention of the birds. They leave the trees, giving the walkers a constant spectacle in the round as they descend. It pleases all the senses.

We can't usually tell who first used some weather word or other, but **#5, petrichor,** is the name that two Australian scientists came up with in the 1960s when they saw that the English language lacked a word for the aromas released when rain hits soil, catches its scent and bounces it into the winds. Like the mist over a flute of prosecco.

They performed something called 'steam distillation' on rocks and found, trapped inside, a yellow oil. In warm weather, plants emit the base of this oil, which then mixes with chemicals emitted by the soil's bacteria. The smell arrives – even if faintly – before the rain, when air is humid and moisture disperses the oils into the air.

Something so primally evocative, if it doesn't already have a name, deserves a pretty good one and our Australians had scope to get poetic.

So they took the old Greek word for rock (*petra*, as in petrified, as in turned to stone) and wedded it to the fancy *ichor* (also Greek: the sweet blood of the gods).

The greens are now greener. The trees, of course, appreciate the kelsher. And the walkers appreciate the trees. They too smell better. They also sound better. Each tree has its own sound in rain. Our weather words on these islands don't include a vocabulary for the music of trees – we lack, say, the Japanese *matsukaze* (松風, the play of wind among pines) – but we know the sounds themselves.

In fact, if our walkers wished to, they could record the rainy sounds around them, loop them and release a very popular sleep-aid podcast. They don't, though. It doesn't occur to them.

They're not listening consciously, but they do notice the change in sounds before they notice that the kelsher is dissipating. They're in the trees. Water falls, not from the clouds. The leaves have held just enough water for a kind of rain that we need a word for: the after-shower that happens if, say, a breeze moves the wet leaves. It's happening now, because some other insects prefer to emerge when rain stops. This prompts more birds to leave and shake the branches. As they emerge from the dripping canopy, the scene becomes dimpsy, in more ways than one.

When the clouds are low and the drizzle is thin even for drizzle, the weather here is **dimpsy (#6)**.

As it happens and as you might expect from the 'dim' bit of 'dimpsy', the 'dimps' are also the dusky twilight, and the sky as daylight fades is a dimpsy one – so to be sure you're not misunderstood, you might use dimpsy to mean only those thin drizzles that sometimes come to us at twilight.

It was over as abruptly as it began. And it hadn't been forecast. Had it? 'Microclimate,' shrugs one of the pair. The other replies with a grin: 'Microclimate.' A word they've heard more often in the last few days than ever before. The woman in the post office who sold them the map: 'Microclimate up there.' The landlord at the bar discussing yesterday's weather: 'Microclimate down here.' So many conversations about the weather and, every time, a pride in the microclimates.

For a meteorologist, a microclimate is a small area – perhaps only 100 metres diameter – that has some combination of shelter and proximity to water which means that its weather is different to the weather in the areas around.

Monty Don, the presenter of *Gardener's World*, for example, happily uses 'microclimates', plural, for different parts of a single garden, since one side of some fence, say, will receive less rain than the other: no small deal for a serious gardener.

A MIZZLE COMES TO DARTMOOR

In everyday speak, 'microclimate' is used much more loosely and often to describe balminess.

Especially in coastal towns, it's used loosely and proudly enough to suggest: 'Feel that warm air. You should spend your summer holidays – and money – here.' In the place we're going to next, it means 'it hardly rains hereabouts'.

We let the wind take us up toward the Bristol Channel. Down to the west we see Bodmin Moor. That isn't where we're headed, but it's important. The clouds round here often come from the south. Not another Spanish plume; of the six air masses that visit these islands, this one here is the Tropical Maritime: air from the Atlantic Ocean bearing water that may have had an earlier existence as part of the warm seas around the Azores, which has cooled just enough in transit to keep its dampness, and arrives here, warm and moist.

As that air mass makes its way inland, it has an opportunity to break up when it crosses high ground like Bodmin Moor; there's less chance it will make it to our destination: Bude.

The highest point on Bodmin Moor is the hill that used to be called Bronn Wennili, the hill of swallows. Brown Willy, though, is the name used by everyone except a few worried dignitaries who sometimes launch mercifully unsuccessful campaigns to have the hill referred to only by its Cornish name. And so the cause of heavy showers Bodmin enjoys remains known as **#7: the Brown Willy effect**.

The air experiences more friction as it passes over land, and the land is closest at Bodmin, where winds converge. As with over Winterborne St Martin, they then stick around; the moist Atlantic air becomes Bodmin rain. And the effect is not felt only in the southwest: in March 2006, they say, the showers did not stop until Oxfordshire's Burford, nearly 150 miles away.

Almost all of the time this landscape can take it: there's somewhere for the noisy rain to go. No such thing as bad weather, only bad drainage. Especially now that the people of these islands spend fewer of their working hours outdoors.

Rain may be the farmer's friend. No rain, no crops. But only up to a point. And the Brown Willy effect may bring what is still, in the West, referred to as **letty weather (#8)**.

Elsewhere, this sense of 'let' is heard only in the clipped tones of the lawyer or parliamentarian, or on a passport:

A MIZZLE COMES TO DARTMOOR

... requires in the name of His Majesty all those whom it may concern to allow the bearer to pass freely without <u>let</u> or hindrance ...

The same 'letting' is happening in 'letty' weather. It's a let; a hindrance. It's too much rain to work.

And that cloud gathering over Bodmin is cloud that won't make it to Bude. Coastal Bude can have almost 800 hours of sunlight in a single summer. Some years, it's the sunniest spot in these isles. Today was always going to be bright in the sense of light, whether or not rain was to come. And the watercolourist looking down on the sea pool is a local and knows that sunniest doesn't mean *no* rain.

She has flask and sketchbook, paints and brushes, a foldable seat – and an umbrella 'just in case of **a fox's wedding**' (**#9**; more plainly termed a **sunshower, #10**, when the supposed opposites, sunshine and rain, are with us at the same time). Quietly, she's hoping for one. Rainbows are popular.

Pick some language or other and you'll probably find in it a phrase for those moments when it's sunny *and* rainy – and the phrase will regularly evoke some animal's imagined nuptials.

In Bulgarian, it's a bear's wedding. In Moroccan French, a wolf's. We have also jackals (Afrikaans), rats (Arabic) and tigers (Korean). Most often, though, the happy event is a fox's.

What business does a fox have getting married? We will never know. The easiest way to coax the phrase into something resembling coherence is to dwell on its longer form: 'a fox's wedding and a monkey's dance'. Meteorologically, the venue has been double-booked and not a lot will make much sense until it's all over.

She unfurls the umbrella, a special one that fastens to her foldable seat, and pours a brew from the flask. Jonas Hanway – born a stretch along the coast – would, if he were here now, look on with mixed feelings.

Jonas Hanway, eighteenth-century philanthropist, was the first man to walk around London holding an umbrella. For this he was pelted with rotten fruit. Londoners were appalled to see an Englishman carrying a pretentious, prissy, downright *French* item that should properly be used only to keep the continental sun off the face of a *mademoiselle*.

Worse was the response from the drivers of hansom cabs, who made much of their trade on rainy days. Worried that the umbrella habit might catch on, they tried to run him down. But they failed to finish the job and, before long, enough cheap umbrellas were made – and bought – that the stigma of carrying one began to fade.

Hanway, then, would be delighted to see the development of the brolly that attaches to a chair. But, in case it's not clear enough how much this man refused to bow to British convention, he would be heart-broken at our artist's choice of strong, milky Typhoo. He was also a lifelong campaigner against the evils of tea.

Today, though, the tea is a comfort. A warm liquid as she sits with her watercolours, sheltered from the warm liquid from above. Puddles gather. The rainbow fails to appear, but a sly thought occurs. She's going to add one anyway. We'll leave her to it, and cross the Bristol Channel to a place where you don't carry an umbrella 'just in case'. We're headed for somewhere with plenty of words for the many times you absolutely *know* it's going to rain . . .

2

RAINING KNIVES AND FORKS IN SNOWDONIA

Over the channel we go, through North Atlantic air. Water gives way to wet land.

In the village below – just like in the villages around – they maintain the jokey practice of naming people after their jobs. 'Hawker the Breakfast', who runs a B&B, introduces her guests to **rain word #11** with her blithe announcement: 'Reckon **Ifan y Glaw**'s visiting today.'

As with people, so is weather given name and trade. On the English side of Offa's Dyke, they talk of Jack Frost; on the Welsh side, the same character goes by the name Jac y Rhew, and he doesn't work alone. He may team up with either of the others in the rhyme in the sampler on the wall of the breakfast room:

MORUS Y GWYNT
AC IFAN Y GLAW
DAFLODD FY NGHAP
I GANOL Y BAW

A translation of this rueful recollection: 'Morris the Wind and **John the Rain** threw my hat into the middle of the dirt.'

This is as good an example as any of the blame unfairly pinned on rain for deeds done by other weathers. It was Morris the Wind who behaved so unkindly. Rain does not tend to remove hats. We will see rain unjustly blamed time and again: sometimes for other misdeeds of wind, other times those of cold.

'If there are clouds on the horizon,' notes our Hawker the Breakfast as she refills mugs and glasses, 'it'll rain this afternoon. Else it'll rain this morning.' It's clear she hasn't finished; there's a deliberate pause.

'Or both,' she adds, grinning.

Raininess as a boast. Across all these isles, many hamlets, many cities and places in between announce with enormous civic pride that they are the 'wettest' or the 'rainiest'. Can they all be right?

Perhaps they can. Some wet village may average the greatest number of days with at least such-and-such a length of time when it's raining. Another damp town may have the greatest overall volume of rain in a given month, or year – or whatever. Others may simply be making it up, or imagining that since they have *much*, they must have *most*; and for those that are entitled to boast, the measurements that prove it come from the simplest of scientific contraptions.

Wider at its top than at its bottom, the rain gauge needs only to sit in its spot, with no pesky branches blocking access to it from the sky, collecting drops to be measured in millimetres. Little use, it's true, in many places outside these isles: in a tropical cyclone, a rain gauge will overflow at best; more likely it will be destroyed. Here in Wales, it's just the job. And the one by the rocky outcrop in Snowdonia known as Crib Goch gets an average annual total of 4,635mm – easier to imagine, perhaps, as five standard umbrellas positioned upright, one above another. At Crib Goch, you can rely on it **bwrw cyllell a ffyrc**, or **raining knives and forks (#12)**.

There's an expression that's even more Welsh than 'raining knives and forks', so Welsh that it's the first suggestion in the British Council's guide to talking Welsh like the Welsh do.

The expression is **mae hi'n bwrw hen wragedd a ffyn**, or **it's raining old women and sticks (#13)**.

Old women *and* sticks. You don't need to be a native speaker to have a sense of whether this describes a passing drizzle or a heavier kind of downpour.

In other parts of the world, there are other phrases that suggest that some kind of cylinder is falling from the clouds. In France, *il pleut des cordes*, it's raining ropes; in Germany, *Draußen regnet es Bindfäden*, it's raining twine; in Spain, *caen chuzos de punta*, it's raining sticks (in this case, spiked sticks).

In yet others, some rain phrases include mammals: in the Netherlands, *het regent kattenjongen*, it's raining kittens; in Colombia, *estan lloviendo maridos*, it's raining husbands.

Only in Wales, it seems, do both cylinders and mammals come at the same time. Old women *and* sticks.

To get to Crib Goch from here, we have to go not just north but up. Up, up, up.

Our route takes us over the Pembrokeshire coast path.

'Do the horses mind the rain?' asks a schoolboy, proceeding at a trot.

'They loves it,' yells their guide. It's true: the horses they've passed in fields have not taken the option of moving under canopy. They've been standing still, blinking, their insulating coats glinting, the wet hair of their heads and hinds taking on a smoky scent.

In, say, London, you might expect 23 inches total rain across a year. In Wales as a whole, it's 55. An exceptionally rainy place, Wales has its own term for 'somewhere you're actually sheltered from rain': **lle diddos, #14**, effectively a 'waterproof place'. Today, it seems everyone is out in it. Not that there's any consistency to the 'it' in question.

In Tonypandy, there's **pistyllio** (rain in the form of a fountain); in Betws-y-Coed, **brasfrwrw** (wide spaced drops); in Blaenau, **a ffestiniog sgrympian** (a short sharp shower).

In Cwmrhydyceirw, you'll have **lluwchlaw** (sheets of rain) – and in Ysbyty Ystwyth? **Chwipio bwrw** (whiplash rain). **Rain words #15 to #19.**

In Wales, the boast of raininess is well earned. As we head north, we see less of the effects of the tropical maritime air mass, that warm, wet arrival from the Atlantic we enjoyed over Dartmoor.

The air mass that calls the meteorological shots as we proceed along the coast is the returning polar maritime. It too is wet and has most effect in autumn and winter. Summer, though, does not deter rain in Wales. Far from it. The heating of the land sends the moist air up, which means showers.

For you, up is the only direction. Welsh hills, Welsh mountains.

On the mountains themselves, some adventurous souls are out today. The walkers – walking here sometimes involves your hands as well as your feet – have heard and felt all the surfaces it's possible to hear and feel underfoot. The carpetty champ of crisp snow. The quieter, looser connection that means the snow is soon to be gone. The satisfying shattering of an iced-over puddle.

Their favourite is when the ground looks like it's been raining pearl barley.

This substance is frozen, but it's not snow. It's not hail. Not exactly. There's a perfect combination of ground temperature and cloud temperature where precipitation becomes, not freezing rain, but a crumbly, rimey crystal. It won't bounce like a hailstone.

In Germany, they call pearl barley *Graupe*. And in English, the word for this barely frozen form is **#20: graupel**.

Further up we go. The top of Wales is as high above sea level as these islands ever reach. Air, which can't go in, and won't keep still, is forced to climb the mountains. As it climbs, it cools and, unable any longer to hold the moisture it has brought, it lets lots of that water go.

If you place side by side a map coloured to show the British Isles' rainiest spots and a map of its highest spots, the colours will be in the same places. We descend from Crib Goch, this darkest of rainy spots, along the coast of the Irish Sea and see some of those that are lighter.

Wales is a land that collects rain: compare even lowland Cardiff (1,203mm annually, under the Welsh average) to England's Ross-on-Wye (764mm). And sometimes the *people* of Wales also collect rain.

On a large scale, hydro-electric schemes keep major towns in fresh water. In the north, we pass over more homes with field drains and shallow wells, their inhabitants free from the vagaries, varying rates and iffy futures of a water company. A mum, below, uses a particle filter because her family prefers their baths to be bracken-free; she considers the minor effort well worth it.

But as she looks in our direction, she frowns. She sees a kind of rain that won't reach her.

Outside of rain, a **virga** (**#21**) is something you see in the notation for the plainsong intoned by Gregorian monks: ℩. A little rod.

In the sky, virga are wet trails that fall from a cloud – perhaps an altocumulus, perhaps a nimbostratus – but which evaporate in warmer or drier air before they get a chance to reach earth.

If they turn back into vapour, colder air might accelerate downwards, which creates turbulence problems for pilots; for those on the ground, virga is rain that won't reach allotments, water butts, potted herbs.

You can be most sure that you're seeing 'virgae' in a red sunset, with a gentle wind curving the tail into an angle or even an arrow. Taken in with the cloud above, the whole thing may resemble a daunting jellyfish of the sky.

If there's already been heavy rain, virgae are a clue that the skies will clear; if they appear *after* clear skies, rain is coming soon.

Those who can't see these particular virgae have seen other signs. In the fields, the surface of the clover looks different. The leaves are lifted towards the clouds. While it would be fanciful to propose that the clover 'knows' it's about to rain, for those who can read a field, it's a matter of simple observation.

Seeing – or hearing. Three fields along, the frogs are out. Do *they* 'know' it's going to rain? In Germany, they're persuaded enough to nickname their meteorologists *Die Wetterfrösche*, weather frogs. Our actual frogs know, surely, that it's their breeding season, and frogkind certainly benefits from rain, making ponds more suitable for the laying of eggs – and right now, you can't miss the noise of the males. Not long now ...

And now, yes: the rain has come and it's not everyone that's out in it. A teenager in a café watches the drops descend on the other side of the window pane. He picks one. If it's the last to reach the bottom, he'll find the bravery to ask her out tonight. He can't tell whether he wants it to go fast or slow.

In her kitchen, a grandmother instructs her visiting charges to get their pennies out. They'll be placing bets on which drops will win the races for at least the next hour.

Best of all, a youngster in a car does no more than watch them descend. There are no stakes, literal or romantic. It's just him and the rain, and the thought never occurs that there might exist anywhere he'd rather be.

When he *does* snap to, and becomes more conscious of what he's doing, he decides that here in the car, waiting for dad to finish his snooker game, really is the best place for watching raindrops. Unlike with double-glazed windows, you can hear as well as see them arrive. So you feel the beats. That said, there is no regular rhythm, like there's no pattern to the way the view out of the window is alternately blurred and magnified.

He draws a raindrop in the condensation. Rather, a 'raindrop'. A raindrop in reality does not resemble the iconic raindrop we all know. In a picture, it comes straight down. In reality, it almost always arrives at an angle. In a picture, it has the same shape as an iconic teardrop; in reality, it starts out as a sphere before air resistance gives it the shape of a jellybean. We rarely see these: if a raindrop grows larger than 6.2mm in its radius, it breaks into smaller ones.

Across this small Merioneth town, a tourist has trouble making small talk. Why is yet another stranger talking to him, he wonders, about the weather? He's from San Jose, where you'd rarely consider weather worth discussing: today's will be the same as was yesterday's.

It will be almost the end of his visit when he twigs that his idea of small talk (what he and the strangers do for a living) breaks a code, puts people on edge. Sadly he will not stay long enough to grasp that a conversation about the weather is not some exchange of information about what has happened or might happen in the sky. It is not a debriefing. It is in every important sense the opposite of a weather report. He cannot read a rain chat.

We chastise ourselves for talking so much about the weather, but we do all live in it.

There is a kind of rain that makes much less of a fuss than most.

Once you know your rain well, you don't talk only of the knives and the old women and the forks and the sticks. The Welsh **rain words #22** and **#23**, **gwlithlaw** and **manlaw**, describe this; everywhere has its own version, or versions.

And across all these isles, there is a common – unforgettable and ancient – phrase. Here it is, in the hoary verse once titled 'Westron Wynde', reprinted in full:

Western wind, when wilt thou blow?
The small rain down can rain, can rain
Christ! If my love were in my arms
And I in my bed again.

The **small rain** (#24) might sometimes be your tears, perhaps; it is also real rain we have long, long welcomed. Here it is again, at the beginning of the biblical book of Deuteronomy:

Give ear, O ye heavens, and I will speak;
and hear, O earth, the words of my mouth.
My doctrine shall drop as the rain,
my speech shall distil as the dew,
as the small rain upon the tender herb,
and as the showers upon the grass.

The phrase is delightfully imprecise: it might be a rain that is small in intensity, in duration, or in diameter of droplet — and it's a phrase that helps us to welcome the small rain itself when it arrives.

We've heading to the north coast now. Geese who appreciate a rainy day whoop gleefully at each other. For the coastal path walkers, the smallness or otherwise of the rain was key to a decision this morning. If it was to be fair: trousers. If it was to rain, another decision: walking trousers or shorts?

The facilities at their Airbnb did not include either clover or frogs to help make a prediction but it did have wifi. Which enables a weather forecast. Which is in turn enabled by information about how the weather has behaved in the past.

Which is knowledge we have not had for long.

George James Symons was a teenager when the great Victorian rain panic set in, during the early 1850s. Where had it gone? The poor in the countryside had to buy water by the bucketload to irrigate crops; less was coming from the sky. But how much less? And why? Nobody knew. There were no measures.

No measures, that is, until the teenage Symons took them. He constructed devices, and kept diaries of what he observed. Aged 17, he went to the Pimlico headquarters of the British Meteorological Society, where they appointed him a fellow. He went on to write to every daily and weekly UK newspaper, seeking like-minded rain-spotters 'of both sexes, all ages, and all classes'.

The recently formed body that would become the Met Office soon gave him a job. Like many a pioneer, his interest was specific: too specific to cover all British weather. His boss thought he was overly interested in rain; he in turn could not believe that nobody else yearned specifically to establish a standard reliable practice for measuring rainfall. So he went solo.

He travelled to the volunteers he had recruited through the ads, who he referred to as his 'authorities', leaving them instruments and charts. Members of the British Rainfall Organisation took daily readings, at 9am. Symons amassed 3,408 reporting stations, funding his life's work through donations.

The figures were collected in Symons' magazine, *British Rainfall*; he published tips alongside on how to protect gauges from such menaces as the leaf-cutter bee. Not content merely to record rain from all parts of the UK from

1860 onwards, he trawled through old ships' logs for data going as far back as 1815, a quarter of a century before he was born.

When the Met Office quite understandably wanted a piece of this industry and invited him to re-join, he opted to 'flatly refuse', explaining that the British Rainfall Organisation's 'esprit du corps would be extinguished'.

To embrace the red tape of an official body, he reckoned, 'would be at the cost of that intelligent independence of thought which so greatly rules the progress of science.'

When he died, there were tributes to his 'power of making friends'.

When the news turns to discussing rain and when it uses the words 'since records began' – it was Symons, with his volunteers, who began those records.

We're now over Anglesey. Below us, a house whose inhabitants found a use of the extra time afforded them in 2020 by lockdown. The BRO's collection of 66,000 pieces of paper – handwritten scrawls typically unreadable by machine – was scanned and posted online by the University of Reading.

Volunteers were invited to spend a little pandemic time entering values so that the Met Office computers could use the vast and otherwise lost dataset to help inform future forecasts. For the project the University of Reading put aside a few months. There were enough volunteers, including those in the house below, that it took a mere sixteen days.

Thank you, then, inhabitants. Next: to the rains of the Irish Sea ... and the other side.

3
A FLIUCH MOURNE MORNING

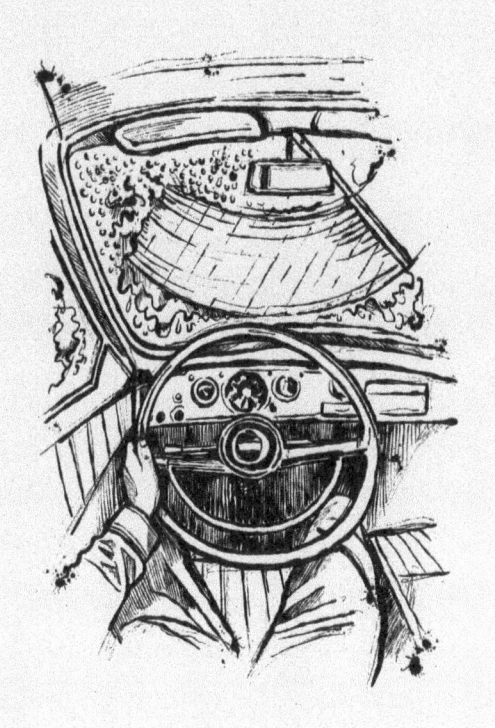

Below us now is Dublin, the city that gave us the frustrated driver James Henry Apjohn, who in 1903 registered a patent for a device that would become ubiquitous. It could be operated by hand or by motor, and could move either up and down or from side to side.

Apjohn's 'Apparatus for Cleaning Carriage, Motor Car and other Windows' was his solution to **rain word #25, breacbháisteach**, or **half-rain**, otherwise translated as the 'occasional rain' that can't be predicted, such as when you set out in your carriage or motor car under skies that are clear – but only for a short while.

At exactly the same time, American motorists were likewise starting to see that they couldn't see. On a New York tram, 36-year-old entrepreneur Mary Anderson watched the driver shifting sleet from his windows with his hands and had a similar idea, which she patented at the same time. (In Germany, it took a little longer and the patent was awarded to, unexpectedly, Queen Victoria's grandson, Price Henry of Prussia.)

The wiper is now, of course, standard, the rhythm of its thud so comforting, the dual arcs it creates so familiar that you might wonder why Anderson failed to drum up interest among car manufacturers. By the time common sense prevailed, her patent had expired.

Many cars, many wipers, many practices. Wedding ribbons on the wipers of newly married couples, left until they come off by themselves to ensure the union receives maximum possible luck. Leaving the wipers at a remove from the glass to deter them from sticking due to heat, or from sticking due to cold.

To the pious man below, the wiper's constant removal of stains is a reminder of his deity's eternal forgiveness. To most of the other drivers on this road, it's a metronome that may or may not match any other rhythms inside the vehicle. Below, in the suburbs, a driver pulls over. Wipers may well be an ingenious invention, she reflects, and these ones do their best, but some rains are too much. This one's a rattling, clanky **pelter** (**#26**, a sudden and noisy, drenching downpour). She gives up fighting it, changes the station. Now the rain keeps time with the music, a new metronome. It's *that* Taylor Swift song. She can only smile. What's coming next? *That* Rihanna one?

In Gaelic, a bell is a 'clag'. In County Kerry, a heavy rain might be called a 'clagar'. In parts of West Cork, though, the word is extended to something more rhythmic and pleasing: **#27, clagarnach**.

Time was, you'd define clagarnach as the sound made as a heavy rain hits a metallic roof. Nowadays, it's more likely to be the sound of rain on a pane of glass. The key thing, though, is *which side* you need to be on to use the word: you're under the roof, or on the dry side of the glass. Inside the car.

It *is* that song by Rihanna – or, to use her fancier title, Ambassador Extraordinary and Plenipotentiary of Barbados, Robyn Rihanna Fenty.

A person's relationship with their umbrella is very different in Barbados. When rain comes, it comes with absolute conviction – then it stops like someone has closed a tap. Everywhere is then instantly hot and everything dries. There are parts of Ireland that feel like they've never been dry.

It's not the same, though, all over this island (and all over its associated smaller islands). Let's turn our attention to the south west. If you were to keep going in that direction, past the coast of County Clare, the next land you'd see would be the Bahamas. But we stay here, downwind of a 4,000-mile stretch of Atlantic, the prevailing wind coming *from* that direction. As, handily, does the Gulf Stream. So while Ireland is as close to the North Pole as, let's say, Newfoundland and Labrador, winters here are so much milder.

Up here is where that warm, moist air is prone to meet colder, drier but denser air from the Arctic. The weather will take it in turns. After a cloudy humid spell with plenty of rain, there's a period that's brighter, colder – and full of showers. The end of the dry period between these is when you're most likely to hear a rueful 'Weather can't make its mind up!'

While that applies across the board, it's even more so in the west. In the east, you can expect it to rain on two days in every five. In the west, it's more like three – at least one of which may be a hefty and committed **bull rain #28**, when you might declare that it's **raining cobblers' knives**:

A FLIUCH MOURNE MORNING

#29, the Irish equivalent of the 'heavy metal things that don't really fall from the sky' that in England is **raining stair-rods** (**#30**).

The land itself is almost as green as the displays in its gift shops.

So often the question about rain is not 'if' but 'which'. No raindrops hit the ground outside a Galway café, but the friends who meet in its doorway both wear raincoats, drops on the lenses of their specs. They'd looked out of their respective windows, seen a similar dearth of splashes and even of puddles, but knew that this did not mean they'd stay dry . . .

A **soft day** (**#31**) is one that might, out of the window, look misty at the wettest – but you should know that when you go out in it, you'll feel it's snuck inside your clothing. It renders the sky and the river and the buildings one colour, and the colour is grey. Rain – at least as it's depicted on weather maps, visible droplets from above – may not come at all. Certainly, there's no point your grabbing an umbrella. This is not rain that comes from any specific direction. It comes from them all.

And since there's equally no point bemoaning a soft day, the words are often followed by 'thank God', from the old, old song that begins 'A soft day, thank God!':

Briar and beech and lime
White elderflower and thyme
And the soaking grass smells sweet
Crushed by my two bare feet

Teas ordered, the friends congratulate each other on being avid 'rain-spotters' (a recurring joke) and chuckle at those passing with the wrong paraphernalia: those who do have an umbrella, say, but don't have a hood. Unluckiest of all: those in denim, the skinnier the fit the more hapless.

'Nothing I hate more than jeans clinging to my thighs, but I will miss soft days.'

'Miss them? You're moving? To where? California? What're you on about?'

'I'm moving nowhere. A man in the newspaper says the summers are getting warmer and the winters wetter, but the soft day? Forget it. Hard days is what we'll be getting.'

They think it through. It's true: you don't now hear people saying 'it's a fine soft day' so often. Phrases die out all the time but with this one it's not so much the phrase that's fading away as the very thing it describes. It's often sad to lose a phrase; it's a tragedy to lose the softness of our weather.

A FLIUCH MOURNE MORNING

'**Fliuch salach!**' they both say at once: that's **#32, filthy wet**; not unlike **#33, fearthainn a rachadh trí chlár darach** – rain so persistent that it would go through a board of oak.

Up the coast, it's drier – but dandelions are closing. A sure sign rain is on its way. Perhaps it will be a **skiffle**, a drizzle moving lightly from place to place (**#34**).

The dandelions are not avoiding rain, not exactly. To continue to describe them as if they are people, they are ... waiting. Some of the water will come in, and open up an appendage on the flower's seeds. It's about 100 bristles, it's called a pappus, and it opens out like an upside-down umbrella. When wind comes, the dandelions will open again and their seeds will travel in a manner not unlike Mary Poppins'. They're thought to be able to travel 60 miles from their initial home, using no energy from their 'parent' plant – hitching a ride instead from the weather.

They will pass over greener and increasingly pleasant pieces of land. Some alight in a **plobán** (**#35**, a pleasingly specific term for a hole in the sod that is first created by the hoof of a horse that later takes enough rainwater to fill it to the brim).

Some will land on the crithir. At heart, a crithir is a spark. Figuratively, it can be a spark of intelligence: 'Is beag an chrithir atá ann', or 'He is not the brightest bulb in the box'. And just as a spark disappears, unreliable land might also be **crithir**: if it's swampy underfoot and about to give way, or if, in the case of **#36**, the land has been ploughed and is drying from rain and taking on a post-wet crumbly quality.

And some of the seeds make it to Aiden's favourite beach. He likes to be first in to the water, as soon after sunrise as possible. And when it's raining? he's asked from time to time. And his answer: the expression *isn't* 'you'll catch your death of wet'. Besides, there's no such thing as wetter than a swimmer.

The sea, Aiden reckons, smells its best first thing in the morning when it's being pounded by raindrops. It's the kind of thing he hears in descriptions of perfumes or oils for diffusers: people apply ocean mist to the room, or spray themselves in coastal rain – but, he says to himself, I'm right here *in* it . . .

The English language does not have its own version of the Dutch coinage 'spookregen', used when an app suggests

A FLIUCH MOURNE MORNING

that it's raining – but it isn't raining. If we do want our own version, though, we should make sure we're sure what the app is trying to say.

Because later, further north, someone is in possession of a fact that his sister will find almost impossible to accept. That fact is that when a weather app or website looks like this ...

1500

☁︎

17°

... it is not saying that there will be rain from 3pm. It is saying that there will be rain *until* 3pm. The sister is about to go through the same stages that everyone experiences when someone tells them this.

First: denial. Is my brother winding me up again? No? Then: this just can't be. When I write under 'May 1' in my diary, it sure is *not* my appointments for the 30th of April. This just ... can't be.

Her next stage is anger. And, in fact, that's the last of the two stages.

It's a scandal, they both conclude, to have an app that gets it so right so much of the time but gives that information in a form almost every user will misunderstand. And an hour can make a difference.

In an hour, it can go from **#37, a dry rain** (an insubstantial rain that may come and go without your noticing) to something noisier and inescapable: what here in the north might be described as **floggin'** (**#38**) and elsewhere as **ag cúr cabáistí** or **raining cabbages here (#39)**.

And so it is here. In an hour, it's **raining upwards (#40)**, though not literally. There are times when the clouds are so dense with water that the drops fall with such determination that, when they stop falling... some of them spread. If they land on a river, they spread out, and the same goes for a pavement that has become a temporary rivulet of sorts.

Some will splash: a windshield is designed to make just that happen. But if they land on what a chemist might call a 'hydrophobic' surface – one that repels water, like many plants' leaves, or some birds' feathers – it's going to be a bounce. And a pavement that has not yet become a rivulet might be a platform for bouncing too.

A FLIUCH MOURNE MORNING

Along this city back street, we see one corner that's changed because of bouncing: a newish building of brick has gone green at its base, where its colour was once – naturally – brick. It's a kind of moss that can live on brick stone, but only if it's fed regularly by rainwater bouncing back onto the wall.

Out on the main shopping street it's a different scene again. A worker on a lunch break thinks: 'I knew it would rain *because* I left my umbrella at home.' This morning, these streets look different.

An artist has got hold of a spray with a quality that is also our **rain word #41: superhydrophobic**. She's made stencils of quotations and sprayed them, alone and at night, onto the pavements of the busiest areas. Now, as rain discolours the ground, the areas she has sprayed remain lighter and shoppers point at the messages revealed. Some of the messages relate to the rain that has revealed them, like this one from 'We Plough the Fields and Scatter':

> . . . *and soft refreshing rain*

This is about as wholesome as art can be. It cleans itself away over time. It costs nothing to see and it is all heart. Less wholesome is one of the mammies, watching the dramatic 'bounce back' effect of a wall now bearing Percy

Shelley's words 'I bring fresh showers for the thirsting flowers', who remarks: 'See that splashback, they should use whatever-that-is on the wall outside the pub, give the urinators a nasty surprise.'

Children pelt past splashily. They call this 'walking in the sky'.

In the suburbs, Derek looks with immense satisfaction at more bouncing drops. 'Down for the day!' he chortles. His last Christmas present to himself was an 'all-weather clothes line'. It resembles a skinny hut. There's a roof, sloped of course, and the 'walls' are slats that likewise direct the drops down and away. It will soon pay for itself in non-use of the tumble drier, he will not tire of telling whoever will listen, and he won't miss the alternative: the smell of cottons drying inside. Best of all, it gives him year-round opportunities to rehearse his bon mot about having a system for his laundry 'that uses the latest in sustainable wind and solar power'.

It's more wind than solar today (and will be tomorrow) but the point stands. Derek takes a snap and captions it '**FORLACHT!**' (**#42**).

The English equivalent of this word is **#43, deluge**, which – just like dilute – means 'washing away'. The Latin word 'lavere' (to wash) is there too in 'antedeluvian', the word to describe the time before Noah built his ark, before a dissatisfied God decided that what our sinful species needed was an unforgettable, world-changing rain. For Derek, though, it's proof that God loves him and wants him to be happy with his canny purchase.

Staying with the Old Testament, it is from a Hebrew phrase meaning the same thing that English has its expression 'the windows of heaven': apertures in the sky through which rain was once presumed to arrive. And from the book of Job, you can take your pick of translations from the Hebrew description of how that rain is stored:

> *Who has the wisdom to count the clouds? Who can tip over the water jars of the heavens?* (New International)
> *Who can number the clouds by wisdom? Or who can tilt the waterskins of the heavens?* (English Standard)
> *Who can number the clouds in wisdom? Or who can stay the bottles of heaven?* (King James)

Further out, near the east coast, a fish plops back into its lake and tries to forget about the strange few seconds it experienced between being caught and being returned.

Fishing in rain, the catcher tells his daughters, is the best fishing. Before you even get to the lake, you have the great joy of spraying an extra layer of waterproofing on your salopettes, not to mention the salopettes themselves...

'... plus, Daddy, you like talking about salopettes because you like saying the word "salopettes".'

'Sure, and who doesn't?'

People will tell you, he goes on, that the fish are scared of rain. Such people, he reckons, have not spent enough time considering things from the point of view of a fish. Not only are fish unafraid of rain; they actively welcome the presence of water, because the alternative is, well...

Even a thunderstorm? Well, we're not talking thunderstorms today. We're talking more of a **craobhmhúr** (#44, a steady quiet sprinkling), which just makes the water a little deeper, and – since light does not pass perfectly through liquid – a little darker at the bottom, and – Daddy likes to imagine – encourages the fish to swim a little closer to the surface.

Does Daddy ever wonder whether he spends too much time thinking from the point of view of a fish?

And: what if he's wrong?

Look, he says.

A FLIUCH MOURNE MORNING

A kingfisher twitches, winks away a raindrop, then pounces and wins supper. Practically built for rain – water-resistant plumage, keen eyesight unimpeded by droplets – if that bird thinks this is decent fishing weather, it really must be decent fishing weather. 'Cause the bird's not a thrower-backer; if *he* doesn't fish, he doesn't eat.'

Soon the birds we hear will be guillemots, gannets and Manx shearwaters; we are heading back over the Irish Sea. We leave behind this place that has such a solid relationship with rain that, in 2019, when the post office commissioned a new 'Stamp for Ireland', it resembled a green field being rained on. Anyone who looked closer at the metallic 'drops' saw that they were the letters spelling out tiny words: **'bucketing'**, **'hammering'**, **'trying to rain'**, **'lashing rain'** and, of course, 'soft day' (**#45** to **#48**, and **#31**)

Time to go.

4

JUST SPITTIN' (A MANCHESTER DOWNPOUR)

Below you, fleetingly, is the Isle of Man, where a reminder – familiar across the rest of the isles – is here expressed in a different rhyme:

My ta'n grian jiarg tra girree eh,
Foddee shiu jerkal rish fliaghey.
('If the sun is red when he rises, you shall expect rain.')

It works in different languages and it works in actual fact.

Today on Man, it's a fox day.

A **fox day** (#49) is one that's bright and clear at the moment when you announce that it's a fox day – but also one that comes during an unsettled period. Rain has fallen and will fall soon. It is not to be trusted. Like, we must presume, a Manx fox.

At the point below us now – the mountain Snaefell, Man's rainiest spot – every walker and railway passenger is thinking and/or talking about the same thing: that from the top, you can see six kingdoms: England, Ireland, Scotland, Wales, Man itself and – it risks sounding cheesy until you see it for yourself – Heaven.

Actually, not quite everyone. Some add a seventh: Manannán, the sea. Likewise, nestled where it is, the weather on Man tends to be an averaged-out version of that of the lands around it: seldom roasting; frosty even less often. Plenty of rain, though. There's room enough between Ireland and Man for the prevailing winds to gather moisture – so Man gets at least its share. **'T'eh ceau fliaghey as cur fliaghey er shen,'** they say here on a fierce day (**'it's raining vengefully'**, **#50**), but the wind keeps things moving, keeps days fox.

Keeps us moving, this time, toward Cumbria's Barrow-in-Furness. Just before we get to the coast proper, you see Piel Island, which was not included in the kingdoms enumerated by Snaefell's visitors, but which has its own crowning ceremony for its sole resident: the King of Piel is whoever happens to run the Ship Inn. Today, the Ship can barely be seen by its visitors, coming in by ferry.

They are two. As they leave the mainland, they declare the day to be **claggy** (**#51**). Claggy means different things

JUST SPITTIN' (A MANCHESTER DOWNPOUR)

in different places: some item of adhesive character might be claggy; something needing a wash might also be claggy – in this case, on this ferry, it means that there's enough damp, low enough, that you can barely see across the water.

A very short time later you can barely see at all. The passengers are in as much cloud down there as you are up here, because they're in **#52: a praecipitatio**. That word, naturally, is related to 'precipitation', a word we will return to soon. The phenomenon itself is reminiscent of one we have recently seen, **#21**, the virga: here, too, there is a cut-off point; here, too, no droplets fall, as rain, from cloud to those final few centimetres that reach the ground. If you're *inside* a praecipitatio, though, it feels nothing like a virga, because the cut-off point is as low as it could ever be.

In Latin, it literally means 'I fall', which is odd, because it's one of the few times that nothing falls in the way we expect. It's that peculiar moment when a supplementary feature extends from the base of the cloud to the ground. A person's first time in a praecipitatio is as memorable and as barely fathomable as their first time in a plane going through a cumulonimbus. Happily, the ferryman has covered almost all of the 1,700-foot crossing before the point where you might consider abandoning the trip. Getting back may be another matter . . .

... and we will not find out how that was resolved as a westerly wind takes us across Morecambe Bay (since winds are named not for the direction they are going but for the direction from which they arrive). This morning, the weather has been what is known here as **#53, dree**. Dree can mean monotonous, monotonously insistent on driving a hard bargain, monotonously long-winded – or, as here, as is probably predictable, monotonously rainy. Lancaster Castle is the closest thing in view, and attractive, especially to lightning strikes – and the increase in strikes in the current climate, which is less monotonous and much more unsettled than when the castle was built, mean that the Duchy is concerned the whole place is in danger ...

... but we stop before that, on the edge of the bay. The day is no longer dree. Now it's truly **comin' deawn full bat (#54)**.

Words with a 'bat' sound have, for centuries, been used to describe things where you feel like you're being hit. Also ones where you are literally being hit: 'lest any bats begin' is 'to make sure there's no fighting'. When a Lancashire rain comes down full bat, it's a rain that could not come any harder. You might prefer to take to your hatches and, well, batten them down.

JUST SPITTIN' (A MANCHESTER DOWNPOUR)

A trio of lunching colleagues look out from the Lighthouse Café at the fat raindrops ricocheting off the bronze pate of the statue of Eric Morecambe, from here a silhouette backed by indistinguishable sea and sky. The two old enough to have watched the 1976 Morecambe and Wise Christmas special are two-thirds of a falafel and hummus salad each into explaining to the younger why they find it funny to see the Eric effigy getting drenched, while avoiding spoilers. 'It's "Singin' in the Rain", but there's no rain,' one offers.

Avoiding spoilers means not explaining that Eric is befuddled by Ernie claiming to be singin' in the rain when there is no rain, then himself becomes 'wet through' thanks to sundry forms of after-rain, like that which has collected in a shopfront canopy.

If there's no rain in the sketch, asks the younger, why is it so hilarious to see the statue getting witchett? YouTube, as so often, gives the quickest answer.

(Technically, the statue cannot be **witchett, #55**, as a person who is witchett has come out in the wrong coat and is soaked through every layer. The alloyed Eric is witchett-resistant and the witchetting is metaphorical.)

Higher now, the closest thing reaches a lot, lot closer to us than Lancaster Castle did. The closest thing is the Pennines. Clouds that approach the peaks from the west are forced up and let out their moisture before the air moves to the other side. They cast **#56**, a **rain shadow**, across the area to the east coast. In some countries, the effect of mountains is spectacular and unmissable. The area in those mountains' rain shadow (the zone on their *leeward* as opposed to their *windward* side) might receive what amounts to no water at all. Dry, sere. Desert.

Here in the British Isles, the effect of a rain shadow is likewise key to our weather, but many many times less dramatic. The parts of *these* isles that lie in a rain shadow get plenty of rain: they may be merely *less* rainy.

On this side, the mountain range finds itself used in a jocular weather forecast. 'If you can see the Pennines from town,' they say around Manchester, 'it'll be raining soon. If you can't see them, that's because it's already raining.'

It's stoical, it's witty, it's wry; often it's even accurate – but not because the Pennines provide a rain shadow, or even that they're mountains. In some parts of the country, the equivalent remark likewise involves high land. 'If you can see Dartmoor,' they say in Devon, 'it'll be raining soon ...'; and further down in Cornwall: 'If you can see the St Austell clay tips, it'll be raining soon ...'

But in Swanage, it's 'If you can see the Isle of Wight ... '; in Edinburgh, 'If you can see the Fife coast ... '; and in Ayrshire, 'If you can see the Isle of Arran ...'

Clouds form imposing architectures. Thomas de Quincey, born on Cross Street in the dead centre of Manchester, declared that nothing seen by his waking eye compared with his visions while eating opium – 'unless in the clouds'. Today's are the kind he had in mind.

On Cross Street now, outside De Quincey House and Wright's chippy, they can't see the Pennines, and the walls shine darkly. Where the water on the road is high enough to reach the kerb, small groups linger well back on the pavements and wait for a break in the traffic that's sloshing the footways. Groups of adults, that is. Once they've gone on their way, children in wellies stomp, thump, stamp and stumble in puddles; some boots are lined and the toes are warm; some are printed with PVC patterns and aren't quite as warm; all of the boots are sealed and completely, reliably, sensibly waterproof.

That said, more children – in smarter shoes – do just the same, drenching their socks. They love a day when the weather's **angin**, **#57**. (Angin can describe a person who is hungover or a piece of food that's less than appetising;

when the weather's angin, it's raining in a way that stops you doing what you had planned.)

One of the kids points at the middle of the road. 'Rainbow!'

His friend announces that he's watched a video that tells you why the flat rainbow is there on the tarmac.

A teaching assistant takes photos of the same effect; the third shot is decent. She has managed to post it, and to look at five or six other people's posts before the splashing on the touchscreen means the phone doesn't know what her fingers are asking it to do. The rain is also attempting to swipe and tap.

Before wiping it on her top (dry under her mac), she turns off the screen so as not to feed it more unintended instructions. She tilts it so that it's facing down to her waist and cranes back her neck so as to be able to see the screen without its being splashed more, but the wipe was not bold enough. She can still see, like a miniature magnifying glass, one drop.

She clicks it back off. Can't they come up with a way of making these things work in rain?

She looks back at the kids. Will the friend, she wonders, convey how, on a rainy day, any oil that vehicles have dripped onto a road will rise up and slosh around *on top* of the water, with which it will never mix? How light will pass down through the oil, get reflected by the water underneath and emerge *back up* from the oil in pleasing spectrum swirls of red, orange, yellow and the rest?

Traditional physics are not playing a large role in the description. The friend describes a bolt of lightning striking a rainbow. The rainbow, he claims, shatters into 'smithereens', strewn 'for miles around'. When a note of scepticism is sounded, he claims to have seen it happen in real life, which clinches it.

And then between her and them passes, in single file, a pac-a-mac family. There's something perfect about this group of six. Each has their own colour, each their own size, including the one in the baby carrier. And you can tell by the crimping in the hoods and the big zips: all the same brand. Properly prepared. She bets they never go out without them. What happens when one of them outgrows one? Does the pac-a-mac wait, compressed in its little carry sack, until a younger member grows in to it?

A rabbi scuttles past, his mind on the words from the prayer:

> *... may it fall as a blessing and not as a curse – may it be for life and not for death ...*

He did not come out with a pac-a-mac, though, and has hastily bought a disposable poncho. Don't bother, mate, thinks the teaching assistant. Those things don't do the job they're meant to. Get yourself a Mackintosh.

And thank God we still have Mackintoshes, she muses, a lifelong devotee. It's a quarter of a century since the coats stopped being unfashionable, just as the factory was about to close. Japanese enthusiasts kept them going and then, finally, there were *Daily Mail* colour spreads.

She tarries longer. The noise recalls the end of the Smiths song 'Well I Wonder', where the band uses rain as an instrument. The pavements and the sides of buildings get inkier as the rain gets stronger. She remembers a phrase her mother used to use: 'it's **real hunch-weather**' (#58); she can hardly fail to – most people's backs are bowed now. No rain is so heavy as to stoop a human being; really it's the cold that's having an effect: we're instinctively tensing our muscles to try and keep them warm. They resemble figures from a canvas by local lad L. S. Lowry, he who announced: 'I can't stand the sun.'

One of the hunchers has made a cape from pages of the *Evening News*. She suspects he has a coat that is good for cold but not for rain – a closer stare confirms that it's wool, and the kind that, once drenched, never sheds its stench. The dogs are not hunchers. They stretch out their necks, astonished at the range and mixture of scents in the damp air.

Living in your screen is impossible in hunch-weather, she reflects, then smiles as she realises that she's become one of those people who have done that thing. She's put her phone away, got lost in the moment. She's enjoying lots of things about one single place, instead of swiping through 20 or so worlds in the course of a minute. She smiles again as she twigs that she won't be able to resist posting about the experience later. (She will omit to mention that she was forced unwillingly into the moment by raindrops on her screen.)

The newspaper-cape wearer darts in to the gallery on Mosley Street, thankful that there remain places where no one expects you to buy something just to be there. He discards his soggy makeshift cape and crosses the rooms. A print catches his eye and he smiles too.

It is by the great caricaturist and friend of Dickens, George Cruikshank. Its title: *Very unpleasant Weather, or The old saying verified 'Raining Cats, Dogs & Pitchforks'!!!*

The saying really is verified: in the etching, cats and dogs literally fall from above. They do not land lightly on the Victorians, who are experiencing hunch weather like they have never known. And the pitchforks have an irritating tendency to land tines-down, trapping limbs.

The wealthiest in the scene look out from their carriage; an enterprising woman already calls: 'cat's meat, dog's meat!'

The real-life rain out there doesn't seem so angin now. But heavy enough that people across town take pleasure in dismissing it as **just spittin'** (#59).

This is a city that does not bemoan its rain. When, in 2008, Manchester United's Serbian defender Nemanja Vidic said . . .

> *In England, they say that Manchester is the city of rain. Its main attraction is considered to be the timetable at the railway station, where trains leave for other, less rainy cities.*

. . . he started right on point, then judged the mood exactly wrong. This is simply not how you talk about it. You don't do rain down, and you certainly don't run away from it. It's not done, Nemanja.

They're fond here of ascribing the great production of cotton in the nineteenth century to the damp air that kept the fibre stretchy (nowadays, machines replicate Mancunian weather inside, whatever's going on outdoors). Certainly, Manchester was helped in becoming the world's first industrial city by water encouraging the rivers to flow energetically, fuelling steam engines. And let's not forget that the rubberised fabric that makes the Mackintosh work so well was made here too – and let's recall the name that the band Oasis started out with (The Rain).

One thing *not* to mention, one that everyone here has at some point seen, but most have suppressed: with about 800mm per year, Manchester is decidedly mid-table, as seen in news stories with headlines like 'The UK's Rainiest Cities – Revealed (And It's Not Where You Think)'. Damp, yes. Frequently overcast? Certainly. But does Manchester out-rain its rivals?

Let's move on.

Beyond the North Sea, in Norway, a chemist has a tale from the 1980s that he likes to tell.

Margaret Thatcher has been invited by the Norwegian prime minister to visit the chemist's research facility. There, she is shown collections of run-of-the-mill rainwater – run-of-the-mill for elsewhere in the continent, that is, with a pH just below seven: very slightly acidic.

By contrast, rain in this part of Norway is considerably more acidic. A chemist herself, Thatcher acknowledges the difference, and the problem – but wonders why he's telling her; put another way, says that, despite what the Norwegians are claiming, the UK has nothing to do with it. The Industrial Revolution started a long way from the fjords.

It is at this moment that they come across a balloon emblazoned with the name of a school in Leeds.

In 1841, the Glaswegian Robert Angus Smith joined the chemical laboratory at the Royal Manchester Institution. His book, *Air and Rain: The Beginnings of a Chemical Climatology*, collects his research, the most striking piece of which was that the precipitation in northern English cities – or at least in the ones where sulphur-rich coal was being burned in industrial quantities – had a decidedly low pH. As he termed it, this was **acid rain (#60)**.

A striking, scary little phrase, one implying that things need to change. So the concept was pooh-poohed by those who were making a fortune from or via the status quo of coal, but it became a household term in the 1970s and '80s: the photographs – the forests with no leaves, just skeletal columns of wood – were hard to ignore.

JUST SPITTIN' (A MANCHESTER DOWNPOUR)

A combination of international treaties, clean-air acts and the forceful intervention of a thoroughly persuaded Margaret Thatcher reduced these particular emissions as the turn of the century approached and, happily, 'acid rain' seems to be a phrase we can now describe as having been chiefly used between the nineteenth and twentieth centuries.

Heading straight northward along the Pennines, the sphagnum moss is coming back. Landscape does well when it's coated with sphagnum, but it doesn't regenerate by itself if the moss is destroyed. And years of sulphur dioxide from acid rain and from coal smoke have destroyed plenty here. An attractive 'bog moss' that ideally takes the form of spongy carpeting, sphagnum has long been described with words like 'hardy' and 'resilient' as well as 'abundant', but it is not invincible.

It holds water well, also in abundance – 20 times its own dry weight kind of abundance. So, not so long ago, sphagnum 'napkins' were used to dress wounds, as it absorbs blood, trading the human fluid for an anti-bacterial contribution to the patient. Further back, it seems, it preserved food well: canny in more ways than one. And so, of course, on a vaster scale, it preserves – or, more recently, preserved – the peat below.

Acid rain transformed this land from literal swathes of green spongy carpeting to **peat hags** (a rain word, **#61**, that marks the absence of rain): bare unmossy peat, alternately drying out and Somme-like. Since it won't regenerate unaided, it's been getting a little help. A sphagnum moss plant is about the size of a 50p coin. Down below, people crouch and chat, planting these by hand, breathing in deeply a seaweedy smell.

Word has spread. 'Sphagnum' trips more easily off more tongues than it once did. Other figures scatter clay pellets resembling the kind of dog food a dog will eat only reluctantly; these contain and protect sphagnum spores. Still more, coming from further afield, strew great handfuls of clumps culled from healthier donor sites. The mushroomy air you get around a carpet of peat has returned.

It's getting greener. We head toward coast.

5

THE SOLWAY FIRTH DISAPPEARS IN SCOTCH MIST

Yet, only two short months had gone, since a man, living on the nearest hill-top overlooking the sea, being blown out of bed at about daybreak by the wind that had begun to strip his roof off, and getting upon a ladder with his nearest neighbour to construct some temporary device for keeping his house over his head, saw from the ladder's elevation as he looked down by chance towards the shore, some dark troubled object close in with the land.

– *The Uncommercial Traveller*, Charles Dickens

The troubled object was the steam clipper *The Royal Charter*; the so-called Royal Charter storm of 1859 was so fierce that Dickens was moved to travel to the coast of the Irish Sea to witness the scene and to talk to those affected.

These waters have seen no starker demonstration of the destructive potential of wind: it reached force 12, the extreme of the then-new Beaufort scale. About 459 were killed; the *Royal Charter* had made it all the way from Melbourne and had less than 50 nautical miles of its journey remaining.

One person indirectly affected was also the first to have used the Beaufort scale, as captain of HMS *Beagle*'s Darwin voyage of the 1830s. Robert FitzRoy had recently set up a small department in the Board of Trade to look at weather: when he watched the Board's barometers plummet that October night, he predicted, among other things, snow in the north of England – correctly. He was extremely concerned.

And when he heard of the drownings FitzRoy was devastated. He had been working on a way to track storms designed precisely to prevent this kind of preventable carnage. Public outrage was enough for the government to give the go-ahead for a new service overseen by FitzRoy – one we now call the Shipping Forecast.

It was not universally popular.

Some had personal reasons for attacking it.

Shipping magnates soon noticed that when storms were forecast captains remained in harbour, wasting time and money – losses that could not, unlike the destruction of an insured ship with all hands, be recouped.

So they denied the science – attempting to predict any kind of future, they insisted, was akin to sorcery, including this new concept of a 'weather forecast' – and leant on the government, who duly withdrew the service.

After more avoidable disasters, the combined efforts of sailors, the newly formed Met Office, and a freshly outraged public brought its return in 1867. FitzRoy's name lives on in the Shipping Forecast area with barely any coast: just a short stretch across the border between Spain and Portugal.

FitzRoy's approach to tracking weather systems proved especially useful on 5 June 1944, the planned date for the D-Day landings during the Second World War.

Eisenhower's people had chosen this date as offering the weather conditions – such as clear skies – that you would need for a surprise raid on occupied France.

The European team tried to persuade him, though, that an observation phoned in by Maureen Sweeney – a 21-year-old postal clerk in Blacksod, the most westerly point in the British Isles and often the first to experience what the Atlantic has to offer – should prompt a rethink and a delay. Another – from a single ship 600 miles west of

Ireland – clinched it for the sixth. This was 4 June. Previously the weather had been as fine as Eisenhower's team reckoned it would be on the 5th.

Happily, Eisenhower was not a sceptic. D-Day was moved to the 6th, from the 5th, which in the end was a night of gales and heavy cloud cover. Had they gone earlier, the operation – the rest of the war – would have been a disaster.

We head up, back over the land. They're looking up towards us again.

'Mountain of cloud up there. Fit to burst.'

'About time.'

As we reach further north, the sky becomes what is here described as **grou**, AKA grim – elsewhere known as **hovering** (**#62** and **#63**) – all of its blue blocked by plump low clouds that look neither patient nor as if they have any other places to be. Threatening rain, or promising it, depending on your circumstances.

Further up the Dales, St Andrew's church is at the top of the hill of the village of Dent. Its south porch is gabled and it used to provide, for those who saw a grou sky as a promise, something more than mere rainfall: **#64, church-lead-water**. Lead here means roof and the water is rainfall which has run off the roof, restorative when sprinkled on the sick – though the healthiest water of all is, of course, specifically that which has run off a Yorkshire church's chancel, due to its proximity to the altar.

Today, one parishioner emerging from the church once the weather has become **slappy** (**#65**, dropping enough water to create puddles underfoot) is thankful to first breathe in the air that's earthier than when she entered and then to hear the noises of boots in puddles, and globs hitting gravel.

Instinctively, faster than her brain can pull up the word 'slappy', or even the word 'rain', she knows what it'll be like when she steps under the eave. Wet. Like her predecessors, the reason she is grateful is to do with health – and she has science on her side too.

Meteorologists use the term **washout** (**#66**) for the tendency much rain has to remove, from the air, solid particles. In this case: pollen. Some of her fellow hay-fever sufferers have been talking about shelling out hundreds on purifiers. When your home's in the Dales, she tells them, you get a purifier for free as many days as not.

Open the window when there's a **sup o' wet** (**#67**) (a sup o' wet which, she invariably and involuntarily adds, will do nowt any harm) and the purifying begins.

(There's actually a little more to it, in her experience: prolonged morning rain – when it's **yukken it down, #68** – has a much more purgative effect than a drizzly afternoon. That said, more is not always better and if there's over four inches of the stuff reported, it actually seems to be able to whip up pollen that had obligingly settled down – right back into the air: what they used to call a **sulphur shower** (**#69**). Usually, she keeps the explanation simple. And of course she's right. The Piriton will stay in its blister pack today.)

Rain or no rain, when she gets home she will empty the contents of her lurgyish daughter's hot-water bottle into the garden. An act she takes pleasure in, like washing veg in a bowl instead of under the tap: deftly avoiding water waste. And a pleasure she finds harder to muster on days when she has read about a water company's tap water needing boiling, or its pollution statistics having been massaged, or any of the rest of it. One person's effort can feel like ... well, like a drop in the ocean.

THE SOLWAY FIRTH DISAPPEARS IN SCOTCH MIST

This morning brought **grow-rain** (#70), a term that's been used by humans hereabouts when they're pleased to see a shower they consider fructifying. If you have an apple tree, or just a set of raised beds that you appreciate, grow-rain is the kind you have in mind when you announce that 'the garden needed it'. The nearby River Eden shows off its spumy **fresh** (#71, the swelling of a waterway when it has been fed by rain from the landscape it runs through). In the older roads, there's a sludgy aroma as the liquid reaches and overflows the brims of the **water dikes** (#72, holes worn in the road that frequently fill with rain and hide themselves, known in colder months as slush-pans when their filling-cum-disguise starts out as snow).

Soon it will be beyond a 'grow-day' and, at the foot of Scafell Pike, in the pub in Wasdale Head – the Wasdale Head Inn, as confirmed by the storey-high black letters, I N N, on its white end wall – one of the regulars inwardly recalls Wordsworth's words depicting 'a torrent bursting, / From a black thunder-cloud, down Scafell's side / To rush and disappear.' The regular will not describe the rain so dramatically himself; nor will he use any rain words with too harsh a sound when spoken. Not his style.

'**Damply** (#73) out there.'

When the weather is mild enough that it actually could be described using a word that starts 'damp...', he'll reach for something else. 'A **donky day**' (#74) if it's mellow and misty along with the rain. '**Parlish slattery**' (#75), with 'parlous' as an intensifier, in the winter sometimes – when it's slushy and muffling. Damply, though, here, equals vast heavy bucketfuls.

It's a way of saying, without saying, 'This? We're used to this. You call this rain?' A weathery locale, we are not far from Britain's only named wind, Cross Fell's Helm Wind. The spots around Scafell have various claims to rainiest-ness.

For the honest little mountain lake Styhead Tarn, the boast is of amassed monthly fall: over four and a half feet in November 2009. At nearby Seathwaite – a Borrowdale hamlet where the road that comes in does not go out again because there's nowhere else to go – the timescales mentioned are shorter; it once got five inches in an hour. In that same November, it got over a foot in a day.

And further up, there's a peak that has long been used as a reference point, the one described by Coleridge as 'ancient Skiddaw, stern and proud'. The piece of local weather lore begins:

THE SOLWAY FIRTH DISAPPEARS IN SCOTCH MIST

When Skiddaw Fell puts on a cap...

...(the cap being the one that Coleridge called 'his helm of cloud')...

...Criffel Hill begins to drap.

Drap (**#76**), or drop. The cloud worn by Skiddaw comes sometimes in clumps, sometimes in triangles. And the forecast provided by the rhyme is for where we are headed next. When the clouds are clear, you can see it from here. We're leaving England again – Criffel Hill is across the Solway Firth.

By the time the River Eden pours into that inlet, it is well fed with sediment. Outflow is not a smooth operation.

The firth holds a mixture of saline from the Irish Sea, fresh water (in the **#71** sense of 'fresh') and silty mixes from rivers like the Eden. The tide acts like a weir, sending the Eden's contribution crashing. All the while, the surface is – the surfaces are – disturbed by more water from above.

Where mist ends and rain begins is hard for anyone to say. There's a phrase that describes this. Or is there?

We use 'rain' and 'mist' and 'fog' without specific dividing lines – unless we're meteorologists, or pilots. There is a technical difference, and it involves measuring not what is there, like how dense with water the air is – but its effect on us.

When you can see through it for more than 1,000 metres, pretty much everyone calls that mist. When the visibility is less than 180 metres, everyone calls that fog.

The zone in between? Aviators still term it fog. Meteorologists call it 'mist' in forecasts for the public and whatever they like among themselves. The rest of us go on feel. If a gust can send it off, it's mist; if it feels like you're walking through a cloud, it's fog, and so on.

Senses similarly shift when it comes to **scotch mist** (or Scottish mist, as it once was, **#77**), which variously means very fine rain, weather that looks like a harmless mist but wets you more than a **sneesl** (**#78**, a shower that won't affect you), and a downpour that the rain-beaten Scots dismiss as mere mist (not at all unlike our earlier damply, **#73**).

THE SOLWAY FIRTH DISAPPEARS IN SCOTCH MIST

It's even hazier when it's a metaphor. Four centuries ago a scotch mist was something that clouded your judgment; now its meaning must be sussed out from context.

For some, it's something that's easy to miss. For others, something that *should* be very obvious is meant in the expression 'What do you think that is, scotch mist?'

For still others, it's a drink with whisky and crushed ice. (In Cockney rhyming slang, it indicates that you've had too much whisky.) And it's probably the whisky that keeps the mist scotch rather than Scottish. The people of Scotland have pretty much succeeded in stopping their neighbours calling them 'Scotch' rather than 'Scottish', so the term lives on only in the broth and the bonnet, the egg and the whisky – and the ubiquitous mist.

On the northern side of the firth, in Annan, a postie looks at the rain landing on the remaining legs of the viaduct that once connected England to Scotland more decisively. Her father talked of the day the dismantling began. He also claimed to have regularly crossed the firth on foot. She mulls on what makes the two nations different, and the same. On both sides of the firth, the weather is the same, and today it is **groff** (**#79**, an adjective describing rain that comes in large drops).

That tropical maritime air mass that we started in has broken up, clashed with others, done its work. Here, we're moved by another that has come across the sea – another 'maritime air mass' – but from a different direction.

The *polar* maritime air mass, while it may be heated a little by the North Atlantic, is not known for its balmy effects.

It may have come from Greenland; today's is from the white top of Canada. It will in time reach the east coast of England. Here, it's felt as north-westerly winds and showers notable for their frequency. You need to know when the dry bits are due.

Whether there will be a **pelsh** (**#80**, a word for a heavy downpour that can also mean a fur or leather coat), say, or whether the weather is to **haizer** (**#81**, to dry up after rain; clothes can also haizer).

That information is now in pretty much everyone's pocket. For some, it is habitually misread in line with some *mis*information. One of the most viral of 2021's 'I was today years old when I learned' memes was a TikTok insisting that a rain forecast of '30 per cent' means not its natural reading (they think it's 30 per cent likely that it's going to rain) but that there is 100 per cent chance of rain, across 30 per cent of 'your area'.

This misreading was shared far beyond native TikTokkers, in family WhatsApps and plain conversations. Minds were blown. Forecasts reinterpreted. Those who read out the per cent as a chance of rain admonished. All wrong.

There have, since the first flurry of noise, been a few rebuttals detailing that '30 per cent chance of rain' means that they think it's 30 per cent likely that it will rain, but none has had the millions upon millions of views of the original.

The changes, in 2011, to the way the British weather is forecast were introduced with noble hopes. We know more, the forecasters said, than we used to: we have more data fed into better programs, so we can predict further into the future and make finer distinctions than was previously possible. So we'll give you more information than we used to and you can make your own, better-informed, decisions. No more would forecasters merely talk vaguely of 'scattered showers'. In came **#82, probability of precipitation.**

It was not an easy segue.

Unreasonable cranks responded along the lines of 'Just do your job and tell us whether it's going to rain or not', which is not a thing that is possible to do.

Even more unreasonably, joyless souls attacked the very phrase 'probability of precipitation'. Why oh why, the droning refrain came, did it not occur to these geniuses to use a simple phrase like 'chance of rain'?

This battle was unwinnable from its first moment. The professionally irritated perpetually believe that they know better – and they suspect, when a boffin uses a phrase of more than a smattering of syllables, that they are being hoodwinked or, worse, patronised. The meteorologist – who has run simulations fine-tuned to distinguish between sleet and sleet showers, hail and hail showers, light and heavy snow, the likelihood of thunder with and without showers ... and drizzle, all of that before we even get to rain – is no more going to use the word 'rain' when talking about 'precipitation' than you are to use the word 'sandwiches' to describe any and all food items.

Besides, it's a pleasure to speak aloud. 'Probability of precipitation.' Especially 'precipitation'.

Around 500 years ago, things that English speakers started to take greater interest in included classical languages and the practice of science. The Latin word 'praeceps' arrived and promptly assumed many meanings. The 'pre', as in 'previous', suggests 'before', and the 'ceps', as in 'per capita', suggests 'head'.

This gave us things like 'precipice' (a headlong fall, or a cliff from which you might have such a fall) and 'precipitation' (hurtling down, or hastily acting in a hurtling sort of a way).

And so rain and sleet and all the rest of it became known as precipitation – because rain and sleet fall on us?

You'd think.

It took a couple more hundreds of years, and the use by chemists of 'precipitate' as a by-product of a chemical process, before weather-watchers began to use a similar term for all those kinds of water – ones which happen to fall, often in a precipitate way, but through pure and pleasing coincidence.

There are many kinds of precipitation – and many kinds of rainy day. Some, to be sure, come with unwelcome wind, darkness or cold; but rain is not the opposite of calm, nor the opposite of lightness, nor still the opposite of warmth. We get gentle rains, bright rains, balmy rains.

In metaphor, rain is typically wielded crudely. It stands for tears, heartbreak. Even Bob Dylan, who knows to give wind rightful stick – an 'Idiot Wind', indeed – repeatedly turns to rain as woe: 'Buckets of Rain'; the intro to 'Make You Feel My Love'; 'A Hard Rain's A-Gonna Fall', in which a forest is sad.

There's a ghastly aptness in the origin of our phrase 'pathetic fallacy', when a writer assigns to things that are not human – often to weather – human emotions.

It was coined by Victorian polymath John Ruskin, who in later life behaved just as would someone we might now describe as suffering from seasonal affective disorder.

He meant the 'fallacy' part. Of the 'cruel, crawling foam' in Charles Kingsley's novel *Alton Locke*, he had this to say, in 1856:

The foam is not cruel, neither does it crawl. The state of mind which attributes to it these characters of a living creature is one in which the reason is unhinged by grief.

THE SOLWAY FIRTH DISAPPEARS IN SCOTCH MIST

It's a tragedy that Ruskin did not recognise when this unhinging happened to him.

He had been a cloud-watcher since boyhood and he well respected a painter or poet who could capture some specific sky or other. In 1884 at the Royal Institution he delivered a pair of lectures titled 'The Storm-Cloud of the Nineteenth Century': he contrasts the weather of his youth with, for example, the 'deep, high, *filthiness* of lurid, yet not sublimely lurid, smoke-cloud; dense manufacturing mist' of the modern world.

As we know from acid rain (**#60**), Ruskin wasn't wrong about manufacturing mist. But as he lost his reason, his diary entries describing the weather behaved, in an attempt to describe reality, just like Kingsley's literary depiction of foam: they were simply not true.

Yesterday utterly pitch black all day, more miserable than I have seen often, even in this evil time.

Fallacious, then – and full of pathos.

One last thing about the percentages in apps: some users glance at '80%' and take it to mean that the rain will be

80 per cent as strong as rain reasonably gets – **seepin'**, you might say (**#83**), likely to make its way through all but suitable clothing. It will be a few more TikToks before we're back on the same page.

6

SPAIRGING IN THE CELTIC RAINFOREST

Below us, on the west-coast isle of Arran, a youthful Nevadan updates her travel blog. Later, when it's all over and she reads it back, she will very quickly realise what her friends and family are appreciating in real time, and choosing not to mention.

'I feel like I'm chasing the rain all round Scotland,' began the previous entry. This one: 'What do you know: I've caught up with my old pal Mr Rain!'

It's raining into her coffee, an effect she is starting to appreciate. It's more drinkable.

Her mistake, understandably, is to think that each time she gets wet, it's what has locally been termed **plenit weather** (#84). When rain descends incredibly heavily and incredibly locally, as if that locality has been fated to become wet by the heavens above, it may 'rain by planets'.

It feels that way to our blogger, but she has no way of knowing that each time she arrives somewhere to find the sky **flobby** (**#85**, swollen, hanging loose, on the verge of bursting), it has also rained the day before she arrived, will rain the day after she moves on and is raining in the areas to the immediate north, south, east and west.

And since each selfie shows her in her treasured Mackintosh, it matters not. A Scottish name for a Scottish coat for Scottish weather. While the Mackintosh was made in Manchester, Macintosh himself – no one knows why the '*k*' happened – was from 20 or so miles inland.

Water in various forms was always important to Charles Macintosh. His father was a dye maker; two key ingredients of the purplish dye called cudbear are lichens, which love the **smuggly** (**#86**, finely misty and drizzly) days provided by Scotland's geography – and ammonia, which Macintosh senior obtained by collecting from the poor of Glasgow what we will call human water.

Like Manchester, Glasgow used its wetness for industry, with dozens of cotton mills dotting the Clyde. Like his dad, Macintosh junior was loath to let waste products go to waste. He encouraged Glasgow Gas Works to sign a deal where, instead of filling vast pits with the rank sludge left behind when they made coal gas, they sold it to Macintosh. This gave him another source of ammonia and, in turn, another unpleasant by-product, one that was all his own.

Naphtha has some downsides. Its smell is sickening. It's incredibly volatile. It was used as a chemical weapon – the napalm of mediaeval warfare – and Pope Innocent II condemned it. But, if you press it between two layers of fabric, and apply enough chemical wizardry to stabilise the naphtha and eliminate its stench, you have yourself the makings of a coat that will keep its wearer dry even in a **bowder** (**#87**, a powerful squally blast of wind and rain simultaneously).

Mackintosh. A Scottish name, as we said. But there's a word that's more Scottish still.

Dreich (**#88**) is the rain word many love to turn to if the topic of weather words should arise. Dreich is a shoo-in whenever the people of Scotland are asked to choose some favourite pieces of vocabulary. It has a life beyond that of a word, though it is also a very useful way of conveying that there's not a lot of light, that there is a lot of rain, and that there's little likelihood of either thing changing any time soon.

Yes, light. If it's dreary out, you're more likely to hear it described as dreich, though 'dreary' is not where the word comes from.

Traditionally, a dreich individual was one in no hurry to repay a debt; you might find your limbs going to sleep during a sermon by a dreich minister. The earlier 'drieg' says nothing about dreariness or the weather at all: its sense is of something to be endured. To 'dree your weird' is to undergo your destiny. A word about stoically getting through life will not have difficulty finding speakers in Scotland.

Change that 'g' on the end of 'drieg' to the 'ch' that (unlike dree, #53) doesn't exist in English English – but that *is* there in 'loch', and indeed in 'och' and in 'Sassenach' – and you have a word that no one has ever needed to be told is distinctly Scottish. Now roll that '*r*' at the beginning, and you've got something that describes not just the wetness of the weather, but a quality of the person speaking it. 'I am Scottish,' it also says, 'and I endure.'

We follow a southerly, up to more islands.

The edges of the Hebrides take the brunt as a weather system arrives by sea. On the beaches, whole rocks are whipped up, hurled about. No animal, human or otherwise, is here to see – or to hear. The roar in the air is terrific.

In a belligerent storm, wind takes the role of chief aggressor. Rain can fill a space with water, but it doesn't tend to knock things down, snap them, move them far from where they've grown, or been built, or been embedded.

A warning is always sounded when wind threatens to become violent. Trees scream. The coy word 'psithurism' has been coined for the gentle sound in woodland when wind makes its way through politely. We don't yet have a word for the cracking claps that trees emit when wind is about to bring them down.

Further in on the island, trees groan louder, cutting through the noise of what is round here called a **thunnerplump** (#89), a sudden, dense, blasting downpour. Here 'plump' means 'arrive with a splash', but if that word ends in too pleasing a sound given that you can't hear yourself for weather, we will turn to the lesser-used but more appropriate variant: **#90**, a **thunnershock**.

The thunnershock becomes lower and fainter in the mix – not because the rain is relenting but because the noise that was present cannot compete with noises coming in. The roar heard at the coast has arrived at the woodland. Beeches strain, birches howl. The birds have gone. An alder topples.

Again, this is not rain's doing. Rain's effect here is on an elm, one that broke years before. It finally fell much more recently and lay more or less unbothered, dark in places, hosting funguses of one kind and another alongside eager beetles. It remained pretty much what you would call a solid object: until tonight. The elm is about to get **blashed (#91**, of a person, plant or hillside: to be persistently battered).

Now, the beating of the water changes the elm's form: the funguses become part of the flow, dispersed across woodland floor. The tree itself is tenderised, readier to return to where it began, under the ground. It looks like pulp, and shines. Lightning and wind are the brutalisers; rain washes and restores life. It feeds the floor.

A few weeks later, a family passes, on a walk they do about once a month. Today is not **blirtie** (#92, full of rain brought in at an angle by the wind). Today is merely #93, **weetichtie** (wettish, in a way that might stop at any moment), with the occasional **skiff** (#94, a short-lived flurry of water), but the time in between has been more **glashtrochie**: **#95**, filled with continuous rain, that brings mud to byroads, leaves them dirty. Here there are no roads, so the mud and the puddles and the soggy residue of elm bark are not dirt on the floor: they *are* the new floor. The boy stomps merrily in the mix. The birds have returned.

We go where the wind goes: towards the mainland. Below us, the water of northwest Scotland known as The Minch. People are trying some dolphin watching. More fanciful people hope for a glimpse of the Blue Men of the Minch, also called storm kelpies, who live beneath the surface and come out when it's raining to quote lines of verse. (When rain out here is not quite a storm, but gusty enough to notice, one explanation that has been floated is a game of shinty among the Blue Men.)

On the mainland now, by the banks of Loch Broom to be precise, a jolly squabble is under way. Yesterday was a **fairies' baking day (#96)**: one that alternates between rain (supposed to help leaven the diminutive creatures' dough) and sunshine (to help it rise).

At issue is whether what happened in the night must have been a **plipe** (**#97**, a heavy, sudden and relatively brief dash of water – and the plumpy sound it makes) or a **blurting** (**#98**, a squally series of short spatters).

— Did you think they gave a name every time a storm came?

— Do they not?

— No, it has to be a big manky one. Not like last night's. Think back, right? The last one was Carolijn, then before that there was Storm Brian—

— It's no' the fiercest name, Storm Brian.

— Storm Brian *did* manage to take out the leccy, mind. And before that was Astrid.

— Whoa, wait a minute. Astrid, Brian, Carolijn. I think I've just cracked like a secret code.

— No' the world's hardest code to crack.

On googling it, they discover that storm names, as well as being alphabetical, alternate between boys' and girls', and that the public can nominate names to the Met Office (and to Met Éireann and KNMI, their Irish and Dutch counterparts). For fun, they plan to submit the name of their friend Una and tell her, if she gets the honour, that it's because of her stormy temper. But there are no storm names beginning Q, U, X, Y or Z.

Different rains as we head further up the west coast. In Ross and Cromarty, a **gandiegow (#99,** a squall where rain comes with a noisy wind, as a gandiegow is also a blustering quarrel); as we approach Sutherland, it's merely **greasy (#100,** a word used to describe a sky that is just on the verge of rain).

Different types of rain, but rain of some kind or another all year round, which is why this coast has long been the site of **#101, Celtic rainforest.** A bioclimatologist assigns the term rainforest to a place that meets three criteria: it has that ever-coming rain; the 'forest' part means not just trees but the mosses, lichens, liverworts and so on that they host – and the temperature doesn't vary too much, such as when you receive a lot of air from the Atlantic.

The Celtic rainforest is psychologically calm, but it's assuredly not quiet. Woodpeckers and red squirrels, in their many active moments, catch the ear, but there are less clattery sounds too. A huge amount of dripping. Lowest in the mix, the gentle plodding of drops on mosses.

A drop that falls here will hit trees, no question. Underneath us is a bank of hazel wood so dense that you would hesitate to describe it as 'trees, plural'. And in the transition from tree to forest floor, there's no clear start or end, soft green carpet rising from ground to coat the bark, as it has since the rains after the Ice Age.

When it hits the hazel, the drop will collect fungal and algal spores from the families living on the trees. Lichen alone has its own ecosystem within the rainforest: life within life. Rain takes spores to find new places, to start new families. Life spreads.

Almost always, the rain we talk about is the stuff heading downwards. Here, though, as elsewhere, there's more than one direction for water. Capillary action draws liquid high into trees. Trees feed the sky. The rainforest takes the rain, and makes the rain.

This clean scene is what you get when rain arrives from the big sea. Today, the rain on the rainforest is **spairging** (**#102**, scattering in multiple, unpredictable directions at once), branches are **sidding** (**#103**, dripping steadily after a fine misting) and red squirrels are scuttling across a **klash** (**#104**, a soft mass of substance saturated by rain).

As we go further north, we'll see fewer trees. Because of the wind, a little, but treelessness is also a choice. Farmsteads and fields. When wind comes to Orkney, there are fewer places to hide.

Earlier today was **#105**, a **roostan hoger**: rain was steady and long, but mainly drizzle. The family on one of the farmsteads are gratified to see it. On the kitchen wall is a woodcut, unsigned – taken to be the work of a great- or great-great grandparent, the wood's origin uncertain – of the shoreline being crossed by a nuckelavee: a demon mostly horse in form that will ravage your crops, that will kill rain. Under the unpretty image, the words:

PRAY FOR RAIN

The nuckelavee, it was said, would not come ashore when there was rain on the islands, even if the day was merely **glaizie, #106**: filled with bright watery sunshine and more rain to come.

The nucklavee resembles some creature of Scandinavian myth. Up north of the mainland, the maritime air mass has often travelled over less sea than its counterparts: colder and thundery. And the words too. They look and sound more Viking. Outside of weather there's Old Norse heritage in 'bairn' (child) – and the Swedish dialectical 'kilta' (to swathe) gives us our word for a distinctly Scottish item of clothing.

The Norse for 'tingle' gives us a **murr, #107**, a passing drizzle. A stronger drizzle is a **hagger (#108)**, but a **daag**

(**#109**, with that distinctive double vowel) might be anything from a fine spray to a heavy shower, with a Norse word for dew in its lineage. It's an old Norse word for dread, too, behind **ug**, a word to describe a sky that looks sure to bring rain (**#110**). Across the rest of the British Isles it lives on, hiding in the word 'ugly', but 'ug' is now solely Scots.

The daughter of the farmstead family looks from the nuckelavee to the window. She opens the front door and listens to the **slob**: the sound of dull, fairly regular splashing outside, **#111**, while she reads the sky and breathes the thick wet air. The sky is **scouthering** (**#112**, suggesting rain is imminent).

'More black weet coming.' She shouts her verdict back inside then shuts herself in. **Black weet** (**#113**) sounds like it's saved for a dark kind of rain, or at least rain from a dark cloud, but it's a distinction from the other kind of wet, 'white weet'. Snow.

Fine. She hadn't wanted to go out yet in any case. This confirms the forecast provided earlier by her squeaking door and ever-so-slightly sticky salt (each a sign of humidity). This, then, confirms that she'll have the couple of hours indoors that she'd hoped for, and it'll probably be finished mid-morning. She'll be out – not as soon as it stops, rather just as it's starting to stop, while there's still some slob to hear.

On her socials, a pal has recently shared another of her life hacks: something about negative ions in the air after a storm with a gaudy array of claims about the effect on those twin modern perennials, the immune system and mental health. Perhaps. She just knows she never regrets a walk when the rain has not quite moved on.

Away north another 90-ish miles – or 80-ish nautical miles, since we're over the Fair Isle Gap – we find land again. Shetland. Here, too, a Scandinavian or other far-off flavour to many words ...

> a **vaandlüb**, a sudden rushing down of rain as if to flood
> a **tümald**, the bursting of a cloud
> an **upslaag**, a welcome arrival of rain on a south wind following a hard frost

... (**#114** to **#116**), alongside one which couldn't sound far-off or Norse if it tried. A **timothy** (**#117**) is a downpour. No one knows why.

In 2018, public bodies offering a weather map were instructed to stop using an inset for Shetland (which had allowed them to zoom in more closely to the rest of the British Isles, to include more land and less sea). 'Nobody puts Shetland in a box' was the demand on the Auld Rock, the idea being that a smaller-scale map was a fair trade-off for a reminder that Shetland is more north, more Norse – more Norn, as is said on Orkney and Shetland – than you might have been tricked into thinking.

Here on Unst, they'll tell you of their record wind speeds: 197mph in the 1990s recorded at RAF Saxa Vord before a gust which was not recorded for the perfectly good reason that it had blasted the meteorological equipment into pieces. There is plenty that's equally stirring, too, in verse. Shetland's 'national' poet, Vagaland, uses variants of rain words **#89** and **#114** in this description of a storm from 'Haem Tochts':

Whin da lift is black wi thunder-lumps
An da vaanloop sokks da laand.

Another, **thunder-wall**, for a dense and towering bank of clouds (**#118**), gives a background to Hugh MacDiarmid's striking depiction of Artemis in 'Sangschaw':

Oot owre the thunner-wa'
She haiks her shinin' breists.

This evening, though, there will be no haiking, no sokking. The sky, vast, is all but empty. The call of the red-throated diver, long and plaintive this morning, is now short and deeper. Here you can, if you wish, take that as a forecast of drier weather – and while it is not technically a goose, you can call the bird by its other name, **#119**: the **rain-gös**.

North of that RAF station, the sole inhabitant of Skaw pauses a walk on the beach and surveys the vastness. He knows well the view along the coast in each direction. The more that's visible, the drier the weather to come. Most days he predicts the time rain will arrive, accurate to the quarter-hour, based on visibility alone. But there is none, now, to predict.

Tomorrow will be a whole dry day. The low-pressure system has passed. Before its next, Shetland will be in a ridge of higher pressure for a spell. No day feels calmer than such a still day here.

Next on our excursion, though: a long, rainy, east-coast stretch.

7
'WE'RE SURE TO HAVE A PLASH' IN NORTHUMBRIA

As we head south, there will be less rain below.

Fife, say. East Lothian. The Moray Firth. All are on the sheltered side of higher land, away from those westerly winds that arrive bearing rain. Much of that rain has fallen already, onto the lands to the west. Here, you might get 700mm in a year, a fifth of the total in the west Highlands.

Less rain absolutely does not mean less talk of rain. It's the uncertainty that drives the chatter, the checking of apps, the specifics of rain words. In the bigger scheme of things, it's the times between the seasons that are most reminiscent of equivalent times past – the first snowdrop, the first conker. We adapt here constantly to changes.

Here, a day may be **shoring** (**#120**): giving the impression that rain may be on its way, from a word which used to mean 'menace'. In Scotland, in fact, the very word **'weather'** can many times mean a rainstorm (**#121**). Weather *is* rain. And so *another* day, one that is **atween wadders (#122)**, is a dry one that follows a wet storm with another due imminently. In many other languages the word for 'weather' is one that means 'time'. Time lurks in our 'tempest'. 'Weather', of Germanic descent, probably used to mean 'wind'; as elsewhere, it's wind that started it, rain that we notice.

Those who live in, say, a monsoon region don't make such announcements as 'it looks like rain'. Either you know it's going to rain today or you know it isn't. Here, we come to know only the bigger cycles. In autumns through to the starts of winters, much of the rain comes from those more persistent Atlantic lows that have made it to the North Sea. In summer, shower clouds are more likely to be created by convection as the ground is warmed by sun. The warm air rises, as warm air does. The higher it reaches, the cooler it gets until gassy water vapour becomes actual water. These showers come and go, falling on smaller areas than the rains we saw earlier borne on fronts or breaking on upland. On any given day: who knows?

In Northumbria, a pair of friends arrange where and when to meet. The first rain word in their exchange will be one that used to be nautical – a twist on 'hoist' – and became 'hoy'.

'WE'RE SURE TO HAVE A PLASH' IN NORTHUMBRIA

To hoy something is not merely to cause it to go up, but to throw or hurl in general, as in hoying money at a problem, and then in a pleasingly complete flip of meaning, it came to denote the actual opposite of its original sense: to come down (**#123**).

> Bring your big coat. It'll be **hoyin' it down whole watter** in about an hour

> OK I will. You're always right

> For some reason

> About the weather, I mean & ONLY about the weather!

> Know my secret?

> OK I'll bite

> I smell it.

> Smell it.

> I can smell a storm coming in

> What are you, one of the X-Men?

Serious. My nostrils get wider

I smell this like, metal

> Are you winding me up?

Same smell as when photocopier at work's been going too long.

> Now I know you're winding me up.

BRB

Looked it up!

It's same chemical in a storm

As in photocopier that's gone too long

Ozone

> I'll paste the sciency bit, hang on
>
> $O_2 + M \rightarrow O + O + M$
> $O + N_2 \rightarrow NO + N$
> $N + O_2 \rightarrow NO + O$
>
> No idea what it means except that I'm right.

> Told you I was always right. And I am.
> It'll be stottin.

Stotting is rain that comes down hard enough to bounce (**#124**, the same word that has since been applied to the leaping of an antelope).

The lightning *is* coming and, when it arrives, it will raise the temperature of the air to 30,000°C. Thunder is the name we give to the sonic boom we hear. What we smell – some of us more keenly than others – is the gas we got to know more about in the 1980s: ozone (which gets its name from the Greek *ozein*, to smell).

Before getting roasted, both the nitrogen and oxygen that make up most of the air get paired up here and there:

lots of N_2s and O_2s. Those bonds get split by the lightning, and while most of the Os manage to pair back up, some do it twice and the air gains O_3, AKA odorous ozone.

Before a storm arrives, downdrafts bring the ozone to the level of people; it moves sideways when it gets to ground level, advance notice of the downpour.

And precipitation is inevitable. Thunder is always, always followed by rain. It's water that's created the charge, and it has to go somewhere. There is a rain word that describes what happens when you get a thunderstorm and don't get wet – **#125, dry lightning** – but rain still falls. It just doesn't reach people. It evaporates. It's really a virga (**#21**) with added dramatic flourishes, drawing attention to what *isn't* happening. Dry for the person using the phrase, sure: but wet above.

More misleading a term is heat lightning – lightning unaccompanied by thunder – which we will not include as a rain word for the compelling reason that it attempts to describe something impossible. If the strike happens far off, and downwind, someone might see it but not hear it. The thunder will still have happened. If a storm breaks in the sky and no one is around to hear it, does it make a sound? Yes, it does.

'WE'RE SURE TO HAVE A PLASH' IN NORTHUMBRIA

We follow the storm, Lindisfarne below us now; then it moves faster than us, down almost to Whitley Bay. On the way we pass over a pier to which an engineer is attaching a sensor. The sensor will use artificial intelligence and what it will sense is sewage, specifically sewage at dangerously high levels. How, the engineer asks himself, has my life come to this?

Then, looking up in our direction, a son with his mother. They stand between swathes of the pavement that have been **plashed** (our onomatopoeic **#126** – they have become homes to puddles following a storm).

— Look, this is exciting.
— Oh, I see.
— No, you don't, Mum. I haven't said what it is yet. Look up. It's that cloud.
— Which one?
— Which one!? The one that's a hulking great tube, like an arch over the sea.
— Sure it's not just that another plane's gone out of Newcastle airport?
— You're kidding, right? It's way too ... dense. And it's new.
— A *new* kind of cloud?

— There was a thing on the news about it.

— Was the date on the news about a new kind of cloud April the first?

— OK, not a new kind of cloud. But there wasn't a name before. That's a volutus. I'm going to share a piccy of it in the local Facebook group.

— You *do* know that the guy who runs the florists will tell you that it's a part of a secret plan to vaccinate us from the skies?

— Christ, Mum, you're right. I won't bother.

A **volutus** (**#127**, whose name suggests the arched ceiling of a vault) sometimes looks as if it's rolling, and it is not connected to the more normal clouds around it. In 2017, its name joined the more familiar cumulus, cirrus and so on that were coined in the early 1800s by someone who couldn't stop looking at the sky.

The teenaged Luke Howard built his own weather station in his parents' garden down in Hackney's Stamford Hill. He would happily have made the weather his life, but as he turned 16, his father had other ideas and sent him to Stockport to live as a bound apprentice to a miserable chemist so devout in his commitment to Quakerism that he wore the unlovely coat and buckled hat that most of his brethren had long before abandoned.

His joyless new life, though, couldn't stop Howard from looking at and thinking about the clouds. They were, after all, there every day, in plain sight. And he pondered which clouds brought which weather. Clouds, he would say, are 'as good visible indications of the operation of [weather], as is the countenance of the state of a person's mind or body.'

In other words, clouds are the sky's facial expression. Reading that expression would be a better business with a collection of words to describe what we're seeing. The words Howard chose were Latin, echoing the trend in zoology and botany to classify and assort the natural world.

- 'Cumulus' means pile, literally a Latin 'mass', as in accumulation.
- 'Stratus' is sheet, from a word for stretching and strewing that also gives us 'street'.
- And 'cirrus' is hair, from a word meaning curl, tuft or fringe.

Three building blocks from which we now have such enjoyable finer distinctions as altostratus and cumulonimbus.

Something so visual cries out for illustration. Photography was impractical for the conclusive reason that it had not been invented, so Howard painted watercolours and, in 1803, published *On the Modification of Clouds*. He kept it simple – 32 pages simple – and so we still now use his terms every day.

Did the world offer unanimous thanks to Howard for this gift?

It did not. Like the more recent grumps we met at rain word **#82**, they began by querying why we would need fancy Latin terms. Could we not have English words for English clouds, some murmured. Upsettingly, their voices were then amplified by a friend of Howard's, who turned against him and proposed a set of replacements: not 'cirrus' but 'curl-cloud'; not 'stratus' but 'fall-cloud', and so on.

All very upsetting. So when a letter arrived, signed Johann Wolfgang Goethe, Howard presumed it was a cruel joke. An intellectual superstar, a polymath, the man with the most celebrated brain in Germany, does not write fan mail to a British manufacturing chemist with a side interest in the weather.

Clearly a hoax, thought Howard, and I will not be made a fool of by giving it any attention. In reality, Goethe had recently tried to do something similar with colours and emotions and, when he saw Howard's work, became extraordinarily excited.

By the time Goethe composed a series of musical poems as an adaptation of *On the Modification of Clouds*, Howard was persuaded that the enthusiasm was genuine. Best of all, Goethe made a persuasive public defence of 'cumulus' and the rest, on the very reasonable and familiar basis that it offers a single set that works across multiple languages and belongs to none.

And so the *International Cloud Atlas*, used by met offices around the world, which followed in 1896, included 'cumulus' and the rest, with very few changes needed. In 2017, some rarer types were added, including volutus and **asperitas** (**#128**, a Latin roughness that lives on in 'exasperate'), where you look up to see something that resembles what you'd normally look down to see – and that is the broad surface of a choppy sea on a turbulent day. You can presume a volutus is at the front of a thunderstorm, and that rain is on its way; likewise, an asperitas does not produce rain, but signals that some is due.

Even finer detail, then, in our categorization of clouds. Other aspects of weather are simpler to measure: temperature is a number; most often what we want to know about wind is likewise numerical: its speed.

Not rain.

With rain, we might or might not want to know how long it will be here (when and whether to go out), what it will do underfoot (which footwear to choose), whether it will more resemble a session in a warm shower or swimming in cold water.

No international system has emerged for classifying kinds of rain; it seems likely that none ever will.

The Irish newspaper *The Journal*'s Daily Edge section did, it's true, suggest that Met Éireann adopt their Fliuch Scale (fliuch, **#32**, meaning either wet or rainy) from level 1 (a grand soft day, **#31**) to 11 (hammering, **#46**) – but they were not exactly entirely serious.

We do well – better – persisting with a mixture of old words and new, some of which depict what rain does, some of which depict how it makes us feel – and some of which shift meaning.

Like the words the children in County Durham's verdant Eden Dene heard their parents use when discussing what to make of the morning's **sea fret** (**#129**, a drizzly, salty haze arriving from the briny). The east coast has a few ways to refer to this frequent visitor, the most pleasing of which is **#130, haar**, with its all-vowel centre. In Old Norse, they said 'hárr' for 'hoary'.

One adult has made the mistake of thinking that because the fret has dissipated, the weather has turned drier with intent. It is only once the youngsters are out of earshot, having been given leave to play outside, that a forecast is checked. 'It's actually aboot to start sylen it doon.'

Sylen (#131) – or siling, to take one of various other spellings – behaves like various Scandinavian words meaning 'to pass through a sieve'. It's not used for rain that comes in fits and starts: this is persistent, as if someone up here has set about pouring water onto those people below. Like Shetland, the east coast is not short of words that resemble Scandinavian ones, or ones which were once Norse. Just as the Vikings crossed these waters, so does some of their weather.

And so the children's plan to build a fort in Eden Dene becomes a more pressing matter. One has brought a tarpaulin (not in anticipation of rain; he has also stowed alongside it such further practical items as a credit-card shaped miniature toolset, an *A–Z of Hartlepool* from the 1980s and a banana-shaped case for protecting bananas that contains no bananas).

They prop up the tarp on sticks as the first drops come, then realise they would like some of it to sit on too. The sticks are rearranged on a groundsheet section secured by stones, and the canopy is ready by the time the water starts to come **evendoon** (**#132**, straightforwardly, without messing about, persistently, pretty much at right angles to the ground and in vast quantities). Among the drops are **thunder-berries** (drops at their heaviest, **#133**).

They laugh and laugh, celebrating their victory over the elements. They lie on their stomachs and watch.

— How come most of the rabbits have gone in but those two haven't?
— 'Cause I don't think they're rabbits. Check out their ears.
— Must be hares. They don't look as miserable and soggy as the rabbits. It's like they've got water-proofs on, standing there preening themselves.
— Yeah, no, they do and d'you know what that extra layer of fur is called?
— No, what?
— A hare dryer.
— Shut up.

— So much for the rain having gone 'cause the sea fret passed.
— Adults are always wrong about the weather. Right. What is it they say, when you're a bairn, that's the sound of the rain?
— Pit-a-pat?
— Pitter-patter?
— Right. And what's the *actual* sound of the rain?

They fall silent. They hear two rain sounds. One is the noise of drops hitting the ground around the area where the hares are still preening, not unlike the sound of a stream rushing by. The other is the drumming on the tarp above them, as it arrives unsteadily through the branches of the shrubs in which they've made shelter, each collision amplified in their little cocoon. And both are sometimes sent to the background by a more pronounced splash of a specific drop hitting wet ground right in front of them.

All these sounds are a long way from pitter-patter.

After an early lunch, when the storm has passed, they stand to get ready to go and surprise the adults with their undrenchedness.

But they have forgotten that raindrops are thigmotactic.

That is to say: the drops followed the law of gravity as they entered the shrubs, but not when they reached the lowest ends of the hundreds of leaves. There, on those tips, drops were to remain either until the sun came out, or until something else appeared from below, disturbing the leaves and allowing gravity to take over again. That something is a group of youngsters who thought the water had gone, now with it running down their faces, down their necks, inside their clothes, chortling and yelling. Those fat drops are cold, but then they were, very recently, ice. And we remember that we still need a word for that kind of after-shower (see **#5**).

Some people feel the rain, as 'King of the Road' songwriter Roger Miller quipped on TV; others just get wet.

They will remember this storm. It's a good thing. There are bad storms too, of course, and it's not always the storms we've enjoyed the most that remain strongest in the memory.

In Whitby, in North Yorkshire, overlooking the North Sea, a good proportion of the visitors are hoping for rain. Here, too, the weather was **raggy (#134)** earlier: the salty, misty drops coming from the sea had a smokiness to them. Here, too, it will soon be **luttering (#135**, raining hard and continuously), which will please the visitors who are here for Dracula. The line in the book ...

at midnight there was a dead calm, a sultry heat, and that prevailing intensity which, on the approach of thunder, affects persons of a sensitive nature

... reminds us that some are more sensitive than others, and our black-clad visitors are certain they're the most sensitive of all. One is also good with numbers. When she spots the lightning, she starts to count off seconds. Seven, eight, nine, ten ... and there comes that thunder. She knows the formula. Divide that count by five: we get two. The storm is just *two* miles away, she tells her friend. And getting closer.

We have rain to thank for *Dracula*, the novel. In 1815, Mount Tambora in the Dutch East Indies spewed about ten cubic miles of nasty matter into the Earth's atmosphere. The atmosphere, quite reasonably, went bananas.

Ash blustered round the globe, temperatures sank, and in far-off Lake Geneva some holidaymakers who had planned to spend their June on and by the water realised that the rain was nowhere near easing off. Feeling cooped up in Villa Diodati and having read a copy of *Fantasmagoriana*, a collection of German ghost stories, the holidaymakers decided to try their own hands at spooky tales.

Since those holidaymakers included Mary Shelley, Lord Byron and Percy Bysshe Shelley, their stories were pretty decent. The best was *Frankenstein*, but another, by their friend John William Polidori, was titled *The Vampyre*, and without Polidori's Lord Ruthven we would not have Stoker's Count Dracula. His is a more appropriate world than most to have been born out of inconceivably heavy rain.

The friends watch from the cliff, dark make-up smearing. The noise of rain landing has changed in the last ten minutes: it's now the rounder sounds it makes when it arrives on ground that's already wet.

Exactly here is the spot where, a coastguard told Bram Stoker when he was holidaying in Whitby, a ship was run aground in the 1885 storm. He saw a dog leap onto shore and run up the church steps, which Stoker re-imagined as one of the forms his vampire might take. Happily, the visibility is just about poor enough that a mind of a sensitive nature can plausibly wonder whether there might be creatures in the water tonight.

8

PLOTHERING AGAIN IN THE MIDLANDS

We will not cross the Pennines. We stay on their leeward side, in the rain shadow, heading at first south-west. Then south-south-west, south-south-east.

South.

York Castle, the River Aire, the M62. A family – full of IKEA meatballs, children arranged around boxes – debates whether to close the sunroof, through which protrudes eight inches of VÅGE storage units.

'It's just a mug . . .' insists Mum, as she accelerates.

Mug (#136) is another word that can as ably describe a mist as it can a moderate drizzle. It is used here to reclassify what's happening – from what everyone knows it is (a drizzle, which is about to get serious), to something that means the car needn't stop for a rearrangement.

'. . . and we don't close for mist.'

'But mist doesn't – drip?'

She's right to put her foot down. The rain will turn: a mug may always turn **dringey** (**#137**, the sense is that we are 'drinking' from the sky); so it will today. Dringey weather is deceptive: it looks like you could nip round to post a birthday card by slipping into a pair of crocs without grabbing a coat. You will return needing a complete change of clothes and a new envelope to replace the one with the now-unreadable address.

Worse – for a driver – the roads will become keechy. Inside, keech is a good thing, depending on your palate. Keech is fat congealed after melting. For some, the best part of a meal is taking a hunk of bread and sopping it through the remaining gravy to the keechy parts of a plate. Outside, nobody appreciates a **keechy** (**#138**) road: one that's greasy and slippery after a shower. When it's the traditional, rural, ground, the traditional term is **#139**: **mudge**, a mix of what is beneath your feet and rain that has just fallen.

Still, no road stays keechy for long. Not so long ago, the same went with another form of transport ...

Leaves on the line doesn't look much like leaves: this is truly a rain phrase (**#140**). For leaves to disrupt a timetable, as well as falling and blowing on to the track, they need to be rained on, then heated by those lucky trains that *do* make it through: heated and heated, mutating and coagulating until they take their final form of tar.

It's sticky, it's stubborn and modern diesel and electric locomotives are not bulky enough to simply squash the grot into irrelevance.

It's no joke – those who have to think about such things call it the black ice of the railways, and the trains' acceleration and braking times are hobbled just as cars' are by actual black ice – but it is treatable. Of late, nineteen MPVs (multipurpose vehicles) have been deployed to nip ahead and pressure-wash a clear path for the rest of us (also reconnecting the electrical components, which let signallers know once more whether a train is there or not). But why only recently?

In central Europe and Scandinavia, trains have long been equipped to deal with much more challenging conditions. In those places, too, pretty much everyone has the right clothes ready for when that weather comes. Our half-hearted approach to dealing with leaves on the line is the same denialism as that combination displayed by some of those living in the British Isles, i.e. not wearing the right coat, then moaning when we end up with wet woollens.

There's also a labelling problem. All this time, if the phenomenon had had a name that didn't suggest that the problem could be resolved in a few minutes with a stout brush – 'storm tar', say, or 'superheated grunge' – the delays might have been more understandable, the whole thing treated with less derision.

The rain isn't going away. Right now it's coming **heavens-hard** (**#141**, when the heavens are as heavy as heavens get).

We head south-west, inland towards the Pennines, the 'backbone of Britain' – but still we will not cross them. Kinder Scout in the Peak District reaches up to us now. Kinder Downfall, the 100-foot waterfall, flaps and thrashes about, the wind forcing it to flow upward, its mist thrown toward the clouds cushioning the plateau. We go south along the Pennine Way until heading south-east again, across Derbyshire. At times, we pass over a low wind, the warming kind that exists only on the edge of a rain shadow.

In Chesterfield, the bells of St Mary and All Saints are tuned to D. The church is the only UK member of the Association of the Twisted Spires of Europe. Today, its bells are **hollow** (**#142**), not just in the obvious sense but also in a very specific one: their sound has the distinct timbre that means those who notice the difference can tell that rain is coming. Wind can be hollow too.

The bells can be heard for a good few miles around. One who hears them, and who does notice the difference, looks up and makes an announcement he never fails to enjoy uttering: '**Black over the back of Bill's mother's.**'

Sometimes it's merely dark, not the full black. Sometimes Bill is Will.

There are as many origin stories for this expression (**#143**) as there are Bills, Wills and Billys in the Midlands. There are plenty of people who will tell you that, long ago, they knew the Bill in question, perhaps even knew his mother by name. This is not possible with some of the most popular candidates as the Bill or Will in question.

The Midlands' most celebrated Will is probably the top claimant, which would make the mother Mary Arden of Stratford-on-Avon. But the runner-up is from nowhere near: the last King of Prussia, Kaiser Wilhelm II. His blustering and unpredictable foreign policy, we're told, is likened to an incoming shower whenever anyone uses this expression, which is often. Enough speculation.

'Black over the back...' is a reliable fixture near the top of lists, with headings like Things You Will Only Understand If You Grew Up In The Midlands, where it rightly belongs, so long as you ignore the fact that, for example, you hear in Scotland that it's 'looking dark over Wullie's mither's'.

Perhaps the important thing, though, isn't identifying Bill, especially as 'out Will's mother's way' is also used to mean 'over there', with no reference whatsoever to rain. It isn't purely a weather phrase. Perhaps the important thing is identifying all those water droplets and ice crystals that stop the light from coming through, making things dark. Or black, if that's your preference. And knowing what that means.

A bright dawn, and a farmer winces. The interview on the radio comes to an end, which she knows means she's about to hear the weather forecast. She remembers the first time that this moment in the morning would become what her family calls her 'morning trigger'.

Four years ago the farm had the least spring rain in fifty years. Day after day: dry. The lambs remained in their sheds; each evening she hoped rain would come and do what rain does to grass: increase its foodiness, get it to grow into something with more nutrients for the animals. A dry spring is costly; it takes months to pay the price.

> *... but not such happy news for all of us ... with that low-pressure system coming over, I'm afraid we are going to have some showers, though hopefully not for long ... and for the rest of you I'm happy to say, it could be dry through to the end of the week ... Some good news for a change!*

She flinches. 'Afraid'. 'Hopefully not for long'. 'Dry to the end of the week'. They're right, these *are* her morning triggers.

We hear a lot of talk these days, she thinks, of 'just-in-time' manufacturing and distribution – how it doesn't take much to disrupt a system based on things broadly remaining the same. Well, here, too, it doesn't take much to disrupt the system, but, year on year, weather isn't remaining the same, broadly or otherwise.

It'll all cost more later this year, she remarks out loud (but to herself because there's nobody to vociferate at this time), when your 'happy news for all of us' is long forgotten. There will be other opportunities. Besides, the more important thing is that *this* farm is in the area that *will* get fed from above. Hopefully it'll be **ollin' it dahn** (**#144**, providing a heavy downpour).

Click.

*May, come she early or come she late,
She'll make the cow to quake*

—French proverb

A **cow-quaker** (**#145**) was once precisely defined. It was rain, in May, accompanied by a cold easterly wind. Using the phrase was a rough bovine equivalent to the human recommendation to hang on to warm clothes ('clouts') throughout that unpredictable month ('Cast ne'er a clout till May be out').

Nowadays, you can use the term 'cow-quaker' so long as it's spring, it's cold and wet, and you can see at least one cow.

Spring can be a time of mixed emotions for a cow. Granted, it is gratifying to be back on a green diet. Hay is all very well, but the fresh stuff is sweeter and what do you know of the concept of 'overgrazing'? You're a cow. On the other hand, the barn was roofed. Your winter was spent out of that season's many precipitations. In spring, a morning that starts out with a warm and pleasant breeze can end up with a nasty cold shower. The one they call a cow-quaker.

Incidentally you might, as a cow, decide to have a sit down before the rain starts in earnest. This is not, as some will eagerly tell you, to keep a patch of grass dry for later. What do you care? It is not to keep warm. It is not, in fact, connected to the coming rain in any way. It is for one of two reasons. Reason one: you felt like sitting down. Reason two: the other cows were sitting down; doing what other cows are doing is something that comes very naturally to a cow.

Britain appears to be the only place where people have imagined a link between cows sitting down (which happens half the time) and rainfall starting (also not an uncommon occurrence).

The mud beneath the cows has historically been called plother, from an old word for babble. But the rain that creates the mud is also called **plother** (#146). Plothering on the plother.

May days may change but the month we think of as least predictable is April. We may be right.

Every April, the jet stream moves north. Those prevailing Atlantic winds abate a little, giving an opportunity for icy northerlies to arrive, just when the seas around us are at their coldest.

Combine this with longer sunnier days and we have warmer ground, heating the air above it. This gives us clouds. And so the old rhyme often turns out to be accurate: **April showers (#147)** indeed bring May flowers.

A bus driver on his day off chuckles at a newspaper feature on 'Politics and rain' as the drops drip from the cover of his tiny terrace.

Black-and-white photographs of polling days: party volunteers deploying umbrellas, ponchos and hatchbacks to escort voters to primary schools and church halls.

Speeches and announcements given alfresco in downpours. The specific sadness of a prohibitively expensive suit glistening like a tacky compère's jacket when no one has thought to bring an umbrella.

There's a short accompanying text. Rain on US voting days helps Republicans, it says, because disaffected

Democrats are more likely to use rain to decide that it's effectively a 'none of the above' day and stay home. It adds that there is no reason whatsoever to infer that what happens in rural America happens here, but concedes that people seem to enjoy parroting that rainy days are 'lousy for left-wingers'.

We should be proud, the article concludes, that each and every time electoral boffins look for a link between rain and turnout, they find, in the UK, none whatsoever. Brits vote while wet.

Elections, the bus driver reflects, are a lot like weather forecasts. Is it more of the same, or time for a change?

Below us now, windsurfers cross Derbyshire's Carsington Water, the most recently built of the UK's proper reservoirs, making it the closest we get to state of the art; it was opened by Elizabeth II – in 1992. We will need many more of these for the floods to come.

Following the path of a westerly wind, green areas come into view. Around Sherwood Forest and beyond, they will tell you that **arkattit**! (**#148**) is an expression inviting you to listen – hark! – to a specific 'it': the sound of rain. Two retired nurses are getting their steps in. And one of them indeed says 'arkattit!' with a trace of a knowing wink.

The other is keen to do just that. She listens, not so much to the rain, as to the distinctive sound of an abundance of rain on its way. This takes the pair to a specific type of conversation. It's the kind where one of them goes on to mention something they find interesting (in this case, the number of words for 'snow' used by people who live around the Arctic Circle) and the other has recently listened to a passionately delivered podcast that discusses the same something at length.

— Apparently, people don't say that any more.
— What, 'Inuit'? But I've only just stopped say—
— No, 'Inuit' is OK. But they don't have fifty words for snow.
— They don't?
— You can make new words willy-nilly in Inuit languages by combining bits, like Lego. So you can take 'snow' and 'flaky thing' and make 'snowflake' ... but you could just as well take *scalp* and 'flaky thing'—
— And end up with dandruff.

— That's where I was headed.
— So how many words for snow does *that* give you?
— Like, millions?
— That's great. I ... I don't get why we're not supposed to say that Inuits have lots of words for snow, though. Especially as they do.
— Ah, this is where the podcast got really intense. It's because back in the day some people said that all those words for snow proved a theory that you can only *see* things that your language has words for. *And* you can disprove this theory if you look at the different words people use for colours. You see, it's all to do—
— But I don't know that theory. Or care about it. It sounds bonkers.
— Fair point, but—
— I just enjoy hearing all the snow words. They're good words, you know?

Outside Lincoln, it's **ambling** (**#149**, starting to rain) in the morning. Rather, starting to rain again. A couple looks out of the bedroom window; neither is sorry to see the skies. It was, after all, his apparent reluctance to go out in last night's **witter** (**#150**, a trickle of rain) that meant that he didn't leave at all. 'You'll get **wetchered**,' she'd said (**#151**, soaked through), though they both knew there was no chance of that; the witter was already easing off.

And now, as the **spitter** (**#152**, the drops of rain making themselves visible on the windowpane), starts to form **gobbles** (**#153**, bubbles created by the arrival of raindrops), it's enough to pronounce the weather to be **suent** (**#154**, marked by consecutive showers), which means no one is going out – not this morning. The curtains are drawn.

We're twelve or so miles upstream on the Trent from the site of the fictional mill on the fictional Floss, near the fictional St Ogg's. In the story, Lucy announces she expects Stephen to make a visit – 'he always does when it's rainy' – and Eliot herself remarks:

And if people happen to be lovers, what can be so delightful, in England, as a rainy morning?

Outside, above them, rain is **tirling** (**#155**, making a sound as it connects with a roof).

In Ruskington, a surveyor – on his day off – notices that some of the effects of Storm Henk's flooding can still be seen on the eevy walls (**eevy, #156**: of the side of a building, damp from precipitation). Being a surveyor who appreciates sweeping Victorian novels, he too thinks of *The Mill on the Floss*. But not of lovers, or of convenient rain. He thinks of 'the careless and the fearful' after the Floss floods, hoping that, when the next inundation arrives, the Floss's banks might . . .

> . . . *break lower down the river when the tide came in with violence, and so the waters would be carried off, without causing more than temporary inconvenience, and losses that would be felt only by the poorer sort, whom charity would relieve.*

The air blows in from that mammoth estuary relieving multiple rivers opening on to the North Sea, the Wash, inhaled with intent by a retiree hovering at her doorstep on the phone.

— You're getting your steps in *now*? But didn't you used to always tell us not to go out in the rain?

— I certainly did not.
— You said you were talking as a doctor *and* as a parent and—
— That must have been your father. I didn't even say not to go out with wet hair. I said not to go out without a *scarf* in the *cold*. So that your airwaves didn't drop five degrees in temperature and kill off your good bacteria.
— You *did* say not to go out with wet hair.
— Only when you were trying to pull off what you called 'post-gym chic'.
— So how come people get colds when it's rainy?
— Probably because they're staying inside breathing in other people's ghastly air, instead of going outside like sensible people.

And Mum goes out. And until this thing that's been going around stops going around, she'll spend as little time as possible in crowded rooms full of the dry air that's happy to hold the droplets expelled in a sneeze for a lot longer than air that has been given a proper clean by the rain. Do these children of mine, she wonders, not remember those long days with the doors and windows open in 2020, on into 2021?

'Get outside', she thinks to herself, 'or you'll catch a cold'.

'Under the weather' is a persuasive way of conveying that you're ailing. You're under, not on top; you've succumbed to natural forces. But it's not a rain phrase. The weather in question is not the raincloud of popular song and trite metaphor.

Sailors, once upon a time, used 'under the weather' more generally to mean 'in a spot of trouble' or, sometimes, 'unduly affected by rum'. In other words, 'gone awry'. And for those sailors, 'weather' meant specifically *wind*: just as mountains may have a rain shadow on their leeward side (see **#56**), so does a boat have a weather side and a calmer lee side.

Again, the wind is the one causing trouble. The rain is not a problem.

As for being **right as rain (#157)**, we have been right as many items. In *The Pickwick Papers*, Bob Sawyer is right as a trivet (then a fireside tripod for a kettle, among other things). In the seventeenth century we were right as my leg. Right as ninepence too: an Elizabethan coin. With ninepence and trivets featuring less in everyday life, that leaves being right as rain. (My leg, for sure, is still with us, but my leg doesn't offer that pleasingly curt alliteration).

There is also, when you're told 'you'll be right as rain by Wednesday', a sense of 'so you'd better get on with things'. No point looking miserable. Much the same as when it rains.

9
THERE'S A BANGE IN THE FENS

Now we reach the Wash – and cross it. Today, as well as the Witham, the Welland, the Nene and the Great Ouse, the Wash is fed from above. It's also fed from eight miles inland: the St Germans pumping station in the Cambridgeshire Fens can shift, to use the recognised unit, an Olympic-size swimming pool in 25 seconds. It's just beyond King's Lynn, England's top port in the era of the Hanseatic League, amid much reclaimed land.

From up here – only from up here – we can discern and follow traces of paths of what were once beds of other, ancient waterways. As the sea rose and fell, fresh channels emerged, draining the marshlands then filling with silt. They were for a long, long time hidden under a thick seam of peat; not any more. Some are less than a foot from sea level. So much for reclaimed land; it's easy to see from here how Cambridge and Lincoln may be the seaside towns of tomorrow and King's Lynn a memory.

The further north you go, the more Norse words you hear. The Vikings didn't just raid Holy Island; they gave us words and practices that culminate in Shetland's festival Up Helly Aa: actual burning galleys and all the rest of it.

On this stretch of the east coast, it's Dutch surnames and Dutch words. In the 1560s, Protestants fleeing persecution arrived, at first known as 'Strangers' and presently part of the fabric of the east of England (many were, in fact, weavers). These waters have for a long time connected rather than divided us people of Northern Europe and many of our words with Low Countries origins have a nautical flavour: skipper, yacht, smuggler, buoy and all the rest of it.

And weather? We can link a local rain term, **bange** (#158), to seafarers from the Netherlands. It's a light rain, but this is East Anglia: there are many words for a light rain. A **dinge** (#159) is a light rain on a dark day. **Lecking time** (#160, as in the Dutch for 'leak') is made up of little showers interspersed with moments of sunshine.

For a rain to be a bange, it needs to be light, it needs to drizzle, and it needs to hang around, really for days. A bange comes with a persistent damp air and a bengy (completely overcast) sky. It may by coincidence leave the flowers below **bangled** (**#161**, drooping from carrying a payload of rainwater).

Across the North Sea, you might use the word 'bange' to mean ... trepidation.

In the East Anglian village below, the wettest summer day of the period from the nineties to 2019 had 43 millimetres of rain. With a 2°C rise in global temperatures, this looks more like 56mm. With 4°C, 61mm. An increase of 40 per cent.

In Cromer, the bange continues and a teenager eyes an umbrella on a kitchen table and tries to decide: What am I eyeing this umbrella with? Suspicion?

No. Certainly not fear. Not even ... trepidation. No, not today. What they said about the things you can control, and knowing the difference? That.

The umbrella is open. Dripping.

Onto newspaper.

The sea level, she thinks to herself: yes, that worries me – but I've decided what I personally am going to do. An umbrella that's open inside? Not going to let it get to me.

Especially as she's recently read how recently the umbrella even became a thing. How can something ancient – even timeless, such as bad luck – have anything to do with this artefact of modern life? At least mirrors and black cats are symbolic, literally iconic. By contrast, what does '☂' mean? Usually 'you're looking at a weather forecast app and it's telling you which coat to take'.

In Ancient Egypt, they said the sun god Ra got in a terrible snit if a parasol was opened inside: creating shade was to be outside only, in full view of the falcon-headed fellow himself. And, our teenager thinks: perhaps this whole thing got started when one person opened a parasol (or a spring-loaded umbrella) in a confined space, threatening to take out an eye – and another person, rather than having to get into the actual likelihood or otherwise of an inadvertent enucleation, pronounced the whole thing Very Bad Luck.

The things you can control; the things you can't. There's nobody else here right now, no eyes to lose. And the poor thing's not going to get dry outside, not in that bange. Plus its spokes are so radiant, so wonderfully symmetrical. There's only one logical response to this unfurled umbrella indoors, she figures.

She seeks out her sketchbook.

☂

Further into the Norfolk Broads, the bange isn't going away. There's been a weather front here: cold air has met warm, and the warm has done what warm does: risen over the cold. And become itself cold. And that air's water has gone from vapour to liquid. A grey cloud. Raindrops. A bange.

Hazy stratus clouds, low in the sky. They bring not short-lived showers but blankets.

A lot of the rain in this part of these islands comes in the summer: solar surface hearing. Land, sea, air, marsh, sky: all often wet.

And it had better stay that way, mutters an inhabitant of Wroxham, peering up from her gardening magazine and into her beloved garden. She has just heard on the radio about the likelihood of future hosepipe bans. Someone is explaining how many years have passed since the UK last built a decent reservoir, how we simply stopped storing water properly when it does come. Heads it's floods, tails it's water shortages. Two sides, the same coin.

It puts our gardener in mind of a prayer she dimly recalls from the same place outside the mosque where they did the Eid rituals as a child. In Medina, Muhammad prays for rain, and it is effective: it rains so hard, and for so long, that he has to come up with another prayer, pleading for it all to stop.

Well, if there is to be a hosepipe ban, she grunts, I hope it duggles right through to the end of the month. To **duggle (#162)** is to rain heavily, unless you're a bunch of puppies, in which case it's to lie all on top of one another. If the two meanings of this ancient dialect word are connected, the reason, alas, is lost.

The forecast had predicted rain today, which hasn't come. She forgives them, though. She read somewhere that if, say, a few trees come down somewhere for whatever reason, there are fewer leaves in the sky, which affects how much water is in there, which is basically what weather is. And you can't predict everything. Besides, the forecasts are a tad more specific nowadays than something else she read somewhere else: the first weather forecast, in the *Times*, from 1861:

General weather probable in the next two days.
In the north: moderate westerly wind; fine.
In the west: moderate south-westerly; fine.
In the south: fresh westerly; fine.

Besides, besides. She's in luck. It *will* duggle for days.

The age of the telegraph changed the way that we talk about weather. If you don't have a technology that travels from forecaster to forecastee faster than the weather itself, you might as well wait for the weather itself to arrive and observe it then.

There are north winds, to be sure, and south winds, but time after time, much of the action in a TV weather map can be seen travelling from the left of the screen to the right.

Named after a French mechanical engineer, the Coriolis effect is the way winds are deflected by the spinning of our planet. It's as if a wind naturally blowing from subtropical high pressure to lows in the north is nudged additionally to the right.

When the people standing next to the maps talk of the UK's 'prevailing southwesterlies', it's this. Look west and you may be looking towards the weather that's on its way.

On the edge of the Breckland heath (one of its Norfolk edges, not one of its Suffolk ones), rain streams down the windows of a camper van. The rods to keep the windows open at a slant aren't up to the job any more, so the panes are propped out with cutlery. Much better to have them open, the air coming in, the drops plummeting off the end and flying away from the vehicle.

The stable-door arrangement is likewise suitable for rainy days (unlike the skylight). The top half open, the bottom closed. It's almost like being outside.

It's **peppering** (**#163**, falling hard) and **thight** (**#164**, dense, when used of rain or of reeds). This rain is decidedly **sluicy: #165**, descending as if a ship canal's gates have been winched open and a lock's worth of pent-up water is enjoying its freedom, but the sound from the radio cuts through.

> *. . . showers, squally at first. Good . . .*
> *. . . wintry showers. Good . . .*
> *. . . thundery showers. Good . . .*

Every time the showers are – apparently – described as 'good', laughter. They're aware that the 'good' probably refers to some other thing (it does: visibility), but the joke works especially well on a bangey day. When the forecast is finished, conversation returns ... to the weather. Specifically, the deep crimson colour that was overhead this morning.

Presently, a balloon darts noisily around the caravan.

A seven-year-old has been struggling to explain to his parents why the rhyme ...

Red sky at night – shepherd's delight
Red sky at morning – shepherd's warning

... actually works. The balloon is supposed to demonstrate the movement of higher-pressure air to where the pressure is lower. The sound of wind, indeed.

He has reminded them that the colour of light is white, and reassured them that despite what their teachers told them many years ago, the sky does not look blue because it reflects the sea – or indeed the other way round.

He has persuaded them that high pressure brings more settled conditions because 'it keeps all the weather in one place', crucially the air's dust.

And that dust decides which colours of the spectrum get scattered. Blue, with the shortest wavelength, doesn't make it to our naked eyes, so the sky *appears* red.

And if you see that red sky in a morning of our spinning world, you're looking east, at the dry weather of the past, saying goodbye to it, and low pressure is coming in its place. It's going to be wet and windy.

Which means this morning's sky means that this bange is going nowhere, which means a wet walk – hopefully among the red deer who they have noticed also stride out in the rain – which means Gore-Tex.

Upwind of here a vegan musician performs at a fiftieth birthday party. Alfresco. He is able to keep playing, when the bange gatecrashes the event, thanks to polytetrafluoroethylene.

As its name strongly suggests, polytetrafluoroethylene (PTFE) is not found in nature. It is a man-made substance. It refuses to be wet. So it can be used as a lubricant and as a non-stick coating.

In 1969, an engineering graduate named Bob Gore, while working for his father's Delaware manufacturing company, became exasperated – so he'd tell it – with the rods of PTFE he was experimenting on and gave one a sudden yank. He had earlier been coaxing it with heat to stretch an extra 10 per cent in length; the yank stretched it about 800 per cent and he had created, for the first time, a new, mysterious, micro-porous, breathable ... *thing*.

As *his* name strongly suggests, this is the origin tale of Gore-Tex. Gore's career subsequently involved the many areas – in medicine, in clothing and elsewhere – in which a use could be found for the thing. Being an American manufacturer, the Gore corporation also spent much of its time in and out of courts arguing over patents, and the University of Delaware is dotted with Gore halls, annexes and laboratories.

Gore, we can be sure, did not foresee the use to which his version of PTFE is being put here.

Two years back, the vegan musician's instrument tore. Reluctant to replace it with one made of the skin of another goat – or indeed a sheep – he plays 'The Wild Mountain Thyme' on his prized set of Gore-Tex bagpipes.

Out east, there have long been many rain phrases describing things you can see or feel that let you know rain is coming. They're often right too.

The sky might show you **mares' tails (#166)**. These are clouds longer and narrower than normal that form in cold sky and do indeed look like the back of a horse. 'Mares' tails and mackerel scales,' they'll tell you, 'make lofty ships to carry low sails.' In other words, sailors who saw mares' tails expected – correctly – an imminent storm.

Darker, roundish, sometimes mistaken for smoke, **water dogs (#167)** float below other clouds. Their traditional interpretation is pleasingly specific: they indicate *more* precipitation: it has rained, and it will rain.

A **Noe ship (#168)** usually appears alone in the sky and, if you squint, you might imagine it resembles the keel of a large boat, albeit one that has capsized. It shares its other name with the boat most associated with rain: Noah's Ark.

The most distinctively Fens sign of all is not a cloud, is not even in the sky. **Sir Roger's blast (#169)** is a surprising and sudden appearance of whirling dust moving across these flat lands. Surprising because this miniature tornado may appear in weather that is otherwise perfectly calm. Alarming if it moves across to some water that you are navigating by boat: you hear a hissing, you see reeds and sedges sway and you act, fast. A Sir Roger's blast can lift a cow into the air – so they'll tell you.

The most prosaic explanation of the name is that it comes from an Icelandic word for whirlwind (*roka*). The least prosaic is that Sir Roger is Beelzebub and the blast comes from his bottom. Way, way back in time, a Sir Roger's blast was taken to be a sign of an imminent thunderstorm, but nuggets of folk wisdom are very capable of correcting themselves, especially when it comes to something as important to a farmer as rain, especially somewhere so tricky to farm.

By 1898, H. Rider Haggard wrote in *A Farmer's Year*:

> *I have often seen these miniature cyclones in Africa but so far as I recollect very seldom in England, and I never yet met anyone who could explain exactly what they are. When they do occur here the labourers say that they portend fine weather.*

Incidentally: in East Anglia, people take a different approach to the earliest question we might ask ourselves concerning how we talk about rain. That question is: when we say 'it's raining', what's the 'it'?

Is 'it' the weather? The sky? The day? The world? This part of it? Or is 'it' just there because a sentence needs someone or something *in* it, doing something? Like when 'it' is late? Linguists cannot agree.

Those who speak other languages are less enigmatic. Elsewhere 'rain comes', they say, or 'rain falls'. Here in the Fens, though, they draw *more* attention to the enigma. The local version of 'it's raining'?

'That's raining.'

It has **rained itself out** (**#170**, eventually stopped after a prolonged blanket of rain). Deeper into the Brecks, a materials scientist navigates her way between the pingo ponds. Her eye is caught by the daisies.

The daisy almost always does just what its name – day's eye – suggests: remains open in the light and gets ready to close at dusk. When it doesn't – when it behaves like night has come before the night is on its way – this means that cool air has urged the bottoms of the petals to budge so that the whole thing curls up. A good thing too: the pollen will stay dry during the coming rain; for best effect, the daisy may droop as well.

One glance at the daisy's eyes is enough; she texts those back at home: GET WASHING ON LINE ☀️🌝. She's right: there is one cumulus cloud, wider than it is tall: that water is staying up there. It looks so cuddlable, you surely wouldn't guess it weighs a million or so tons.

Presently, she reaches the line of pines planted 200 years ago, rising above and sheltering the Brecks. Their cones, too, are moisture-sensitive. In their case, it's fibres. When air gives a signal that rain is coming, the fibres tug, and inward go the bracts, inward the ligneous scales; the cone is closed for business and its seeds are staying put. As with the dandelions across the Irish Sea, this process demands no input of energy.

Once the rain has passed, when that Christmassy smell has dissipated, the fibres bend the scales outward – people that notice can use it as a forecast – and the seeds are free to join the pollen, the dust and everything else that gets added to the air and stay suspended there until it's cleaned by the arrival of the next downpour.

Our materials scientist has a mental flash. What if a fabric did the same? What if, instead of one coat that's completely waterproof (because it seals you in) and another that's breathable (so you're not being poached), you had one that's made like a pine cone, using hydrophobic and hydrophilic polymers to switch states? Truly, a coat for all seasons.

Today, the cones were open. Tomorrow, they will be sealed to the point of apparent impenetrability and smoothness.

Further south, a cloudspotter and his daughter gaze in our direction. They eye a cloud that, while greying, has a flat bottom and decent visibility underneath. Not the kind of cloud that will bring rain. They also eye an aeroplane.

— So how high up is that, Dad?
— The plane or the cloud?
— Uh, I meant the cloud, but which is higher?
— They're both about two miles.
— So why did you ask which I meant?
— I wanted to know which you wanted to know.
— But— never mind. So our balloon will be further up than them?
— You bet, like two or three times further. Just before the ozone layer. The satellites that it'll talk to are like 75 miles, way out.

They're taking it slowly; eking it out, really. They could have bought a weather balloon rather than assembling one, but the point is to take time, to spend time. Waiting for components to arrive from Turkish online shops. Debating solar cells versus tiny batteries. To include a minuscule camera or not? (Not.) The key difference between 1.8 grammes and 1.9 grammes. Which weather server to share their data with. Most importantly, the balloon's name (the daughter is pitching for Gertie). And hopefully, if it isn't done for through bad luck by a pylon, or deliberately by a military jet, it will waft around for months.

The plan was almost derailed when they considered what would happen if Gertie came down somewhere they couldn't get to. This would be an act of extremely slow, long-range (and, it must be said, minor) littering. Instead of a no-go, though, they're more than offsetting even the potential littering with another new hobby, one that involves ordering long-stemmed grabbers online. Litter collecting interspersed with jogging. Plogging. And if they weren't already feeling they were doing enough 'for the good of the lodge', there's also satisfaction in contributing some genuine weather data to the worldwide pool. Not a lot, it's true, but data gained from the sky is rarer and more useful than the kind you get at the surface.

We've been sending balloons up for some time, though they weren't always so cheaply assembled. In the 1980s, you might find a fallen balloon bearing a chunky transmitter known as a radiosonde and be tempted to return it to a given address by the words '£2 reward, British nationals only'.

There will be no address, no promised reward on Gertie. Gertie will read, simply: Gertie. Soon, she will rise to where we are, and pass us. As she will do later, we will now head south.

10

AND IN KENT, MORE FOLKESTONE GIRLS THAN EVER

We're near the bottom of these islands now, closer to the point where we started. And it's the first rosé vineyard we've passed.

The southeast is the area with the most land between it and the paths of most of those depressions from the Atlantic. Less of their wind, clouds and rain make it here. This is not to say that this locale misses out on rain: that vast capital city generates an extraordinary amount of heat, encouraging still more of that convection we've been encountering out east.

The vines below us like to drink about 500mm of rain every year, if the grapes are not to lose weight: winemakers like this to arrive predictably, and at the right time. In their bottles: the product of this rain on these vines, when also fed by the sun, for you to taste later. Rain as a celebration.

In parts of Kent, if a rainfall is light, if it comes just when you need it most, it's known as a **shatter** (**#171**, a timely sprinkling).

This shatter agitates molecules in the earth and the air carries the scent of **geosmin** (**#172**, a contributor to petrichor, #5). We are very attuned to picking up the scent of geosmin – like, ten parts per trillion attuned – and no one knows why. But it is surely what Longfellow was inhaling before he wrote, in 'Rain in Summer', of:

. . . the vapors that arise
From the well-watered and smoking soil.

As this shatter continues, other aromas join the geosmin. Many of the chemicals we find pleasant to smell in plants are first made in the hairs of their leaves, released when those hairs are bashed by raindrops.

What rain does to plant material, they say, is like a pestle on a mortar's dried herbs and spices. There's a little breaking, then everything smells better.

The shatter prompts a vicar and his family to debate whether it's worth getting their cagoules from the bigger rucksack. The mix of the vines and rain remind the vicar of something.

For years, he has intended to – perhaps close to St Swithin's Day? – use in a service a tale from Genesis 9. Everyone knows about his ark, he might say, but what did Noah do *after* the rain had drained away?

He goes back once more over the events. Noah plants a vineyard. The vineyard produces wine. Noah drinks too much of the wine and passes out, naked. Noah's son, Ham, alerts his brothers to what is going on. A furious Noah casts a vicious curse – not even on Ham but on Ham's own son, Canaan.

It would certainly get people thinking, he's sure of that. It's just . . . he honestly can't make the slightest sense of what the book of Genesis is trying to convey to mankind with this sad and frankly muddled series of half-events.

Exodus, he reflects, is more on-the-nose than Genesis. You just know where you are with a Plague of Egypt . . .

There was a good and honest time for the phrase **blood rain (#173)**, but that time has passed.

Before the good and honest time, people went quite reasonably berserk when it looked like the heavens were hurling blood at them.

In the *Iliad*, Hera advises Zeus not to interfere to save his child, Sarpedon, from his destiny. Since Sarpedon's destiny is being slain by Achilles' friend Patroclus, Zeus's response is suitably dramatic.

She spoke, and the father of gods and men heeded her words. But he let loose on the Earth a volley of blood so as to honour his dear darling son who was to die far from his home, at Patroclus's hands, on Troy's rich soil.

Other tales of similar vintage from around the world have similar scarlet downpours. Back then – and especially if some other misfortune had recently happened or was simply feared – it would take the soberest of sensibilities not to feel apocalyptic when a real red rain came.

After all, the actual explanation – that an air mass has picked up, thousands of miles away, in a place with a geography you would struggle to imagine, desert dust, and blown it into your air, where it has been picked up by what would otherwise be perfectly familiar rain – that explanation is likely to be beyond your ken.

The brief honest period was when we knew just that. Saharan dust storms, say, plus southerly winds.

Once the term was known to headline writers, we adopted a new, tedious pattern: the current one, that goes like this:

- Forecasters announce that some dust from northern Africa is on its way here
- News websites howl that UK TO BE PELTED BY BLOOD RAIN
- A few car-owners in East Sussex need to wash a yellow-brownish film from their vehicles
- The end

There will be no red rain today, happily. Rain, but not red. Soon enough it's going **to power** (**#174**, to descend in torrents); the local spin on the metaphor we've met so many times is **#175**: to **rain 'atchuts an' duckuts** (hatchets and sickles).

And red, but not rain. The red stripes below us are where an arable weed has successfully propagated itself at the feet of hedgerows.

A curious flower, scarlet pimpernel. It has inspired enough affection to have given its name to a gallant fictional swashbuckler and to have been included by Tolkien in the apparent flora of Middle Earth – despite being an invasive species that, when it invades its way to pasture, proves horrendously toxic/narcotic/fatal.

But it remains harmless and indeed looks pretty by the side of a road and, despite having a lovable name in 'scarlet pimpernel', it's one of those that go by many. For the grandfather peering at the side of this particular road, it's a reminder of his own grandfather, who called it 'the poor man's weather-glass' and encouraged study of it to predict the weather, a habit the younger of the pair has kept to this day. Though, over the past few decades of study, he's started to regard the instrument as oversensitive: it seems to close even for an overcast sky, not specifically when it's about to be **coming down hasty (#176**, bringing a noisy, powerful downpour).

But, no one he knows talks about 'weather-glasses' any more; they barely even use the modern-sounding 'barometer', so there's something he doesn't want to lose about this name, the poor man's substitute for an often-beautiful contraption, itself now superseded by an app.

A welcome addition to your company, a group of **Folkestone girls (#177)**? From the name, not unreasonable to presume so. However, these are vast black clouds that fill a sky which, the last time you checked, was blue. Rain you

AND IN KENT, MORE FOLKESTONE GIRLS THAN EVER

hadn't expected, that you have not been able to plan around – like a bolt from the blue – is nobody's favourite kind. A comment in Eleanor Grace O'Reilly's Victorian novel *Reed Farm* leaves us in no doubt:

> *'Like all the lasses as ever I heard on, they're up to mischief,' said Caleb.*

Also known as Folkestone lasses and Folkestone washerwomen, we may never know who decided that these gloomy gamechangers needed, in every form of the phrase, to be female.

We approach that metropolis that causes that extra heat that brings that convection rain. Out East, the rhyming slang for 'rain' is effectively unlimited, since so many phrases end with that sound. That said, there's a preference for one rhyme – **ache and pain, pleasure and pain, Cynthia Payne** – while the most pleasing is the bafflingly nonsensical **duke of Spain (#178 to #181)**.

Below us now is Highbury Terrace, location for the wet kiss in *Four Weddings and a Funeral*, the one where Carrie tells Charles that she had failed to notice the weather – as if the scene is romantic *despite* the rain, when of course the rain intensifies the romance, just as it does in *Breakfast at Tiffany's, Sliding Doors, Spider-Man, The Quiet Man* and a thousand Bollywood pictures where a downpour can make any clothing look like very little clothing at all.

The great cinematic kisses in the baking heat: they come less easily to mind.

Closer to the city centre, a boutique hotel whose bathrooms promise 'the authentic feel of rain'.

Closer still to the river, the site of the old Whitehall Palace, where Clive Owen's Raleigh hurls his fancy new cloak across a puddle to flirt with Cate Blanchett's Elizabeth, though the real-life Raleigh's charms were not powerful enough to prevent the real Elizabeth from imprisoning him in the Tower two miles downstream.

Between that Tower and here is the site of the old Holborn Bridge, where Jonathan Swift ends his wonderful *A Description of a City Shower*:

Drowned puppies, stinking sprats, all drenched in mud,
Dead cats, and turnip tops, come tumbling down the flood.

In his grim vision, then, it is quite literally **raining cats and dogs (#182)** as well as fish and root vegetables. It's a striking enough image that many have been tempted to wonder: is this a real phenomenon – and in that case, is this where the expression comes from?

Did heavy rains bring dogs and cats tumbling down often enough for a phrase to be coined?

People are prepared to go to enormous lengths in hopeless pursuit of explanations for likable words and phrases; 'raining cats and dogs' is as good as any of an example.

Some have somehow persuaded themselves that we should take Swift literally; they have then declared that during storms, some animals – cats, and dogs; it is never explained why it should be solely these animals – repaired to thatched roofs of buildings during heavy downpours.

Chief among this theory's problems is the utter effectiveness of thatched roofs: if the creatures are inside the thatch, they will stay there, completely dry, as that's what roofs do. (If, for some reason, they were to pop outside into the torrent, this is not a mistake they would make a habit of repeating.)

Others have taken an approach that sounds clever but isn't. The sixteenth-century word 'catadupe' – a waterfall – has been paraded as a possible source; there is no evidence for any link and 'dupe' absolutely does not sound like 'dog'; closer in sound is the supposed Greek phrase 'cata doxa', which, we're told, means 'contrary to experience or belief'. A lot of rain, though, is absolutely contrary to neither.

Or perhaps the cats and dogs really 'rained', but not from thatched roofs? Perhaps we should leave the phrase alone. We've just seen it raining hatchets and sickles (**#175**) and near the start of our travels it was raining old women and sticks (**#13**). So many metaphors, in so many languages: it's a bit much to imagine that this single one (**#182**), out of all of them, is based on a rain that really used to happen.

We follow the Thames upstream, raindrops feeding the river, directly from the skies and indirectly through pipes.

A group of Japanese executives on a business trip remark that the weather is like back home. It's true, and it's true of very few other places besides. Japan is positioned not unlike the British Isles. Its 'Europe' (as in, the big landmass to one side of it) is East Asia, and it's to the west rather than the east. Its Atlantic (the vast ocean over which travels air full of moisture) is the Pacific, and to the east rather than the west. The directions are reversed but the effect is similar enough to give a similar variability in rainfall. One difference, though, is that unlike in Tokyo, there is not a culture where you take an umbrella you see outside a shop and leave it somewhere else for someone else later – as Mister Nokajima is about to discover.

The oblongs of green below are Wimbledon. Friends of those patrolling the grounds today – checking the meteorological kit, or on their hands and knees, literally counting blades in a quadrant before bouncing a test ball – joke that the non-July months must be their favourite time of year: when there are no matches. That's not true, though. Well, not quite.

One thing overlooked by almost everyone who visits but does not work here is the Digit.

This is another oblong, but this time small and cut into the green of the ivy, high up on Centre Court's south wall, with equally unobtrusive partners all around the grounds. It shows one of those numbers made up of some combination of seven bars, like a digital watch of the 1980s. It is both inconspicuous and the absolute focus of the attention of the staff who know well the meaning of each of the eight possibilities for the Digit, paraphrased here:

1: Showers now possible
2: Chair umpire *may* call off match, at his or her discretion

3: Same umpire can order a very fast deployment of tarpaulin
4: Like 3, but, like, *right now*
5: The rain is not going to be a brief shower: inflate the tarpaulin . . .
6: . . . and now deflate
7: Roll up, with effort, the heavy wet tarpaulin
8: Start getting ready for some more tennis: look lively, we need a net, players need towels . . .

The system's shape is pleasing. It looks forward. There's much to admire in weather's other number systems – Beaufort's 0 for no wind to 12 for a hurricane – but there is something positive in the way the highest value of the Digit tells those in the know that play may resume.

Given that a rainless Wimbledon comes on average once every couple of decades, the ballet of the unfurling is as much a part of the event as the barley water. We find it so mesmerising partly due to its efficiency: the thing is 8,000 square feet and they get it out deftly enough that often the grass remains so dry that it needs to be watered at the end of a rainy day.

We will overlook the incident in 2016, when a member of the ground staff lost his footing and his colleagues continued to pull so seamlessly and in such practised fashion that he became a hapless, helpless wriggling lump underneath.

And the tarp is easier to read than the roof. The roof appears to do magic, but cannot. It doesn't make the outside inside, or grass behave like clay. It takes time to restore humidity. Play may rightly be suspended even when the roof is closed. A tarp, though, is a tarp. Everyone instinctively understands the rudimentary physics of a tarp.

The Digit, then, is the second-best piece of Wimbledon weather kit. So which is the best? It is not the radar, though the radar is undoubtedly impressive in its detection of 'rain cells' from an hour away and in the fineness of the distinctions it provides every 60 seconds, letting play continue at one end of these grounds while the others get covered.

The best piece is on the roof. It's another oblong, alternately referred to as a bucket and as a tray, clamped in case wind comes instead of or along with rain. A camera points at it, and it is home to a cheerful yellow plastic duck.

Because while everyone involved in Wimbledon appreciates knowing that rain is on its way, which end of the grounds it will reach first and all the rest of it, nothing – nothing at all – is better than this bucket for answering the questions: is it raining here now, and how big are the drops?

Inevitably, it has been named John Quackenroe.

And say you *could* stop the rain. Say cloud seeding was effective: the efficacy they aspire to in Dubai and in Jakarta. Say you could use it to make that rain break elsewhere: send planes to hover over Pas-de-Calais or the Kent coast or wherever, spraying salts so that rain fell there and not here – would you? So that a quarter-final never has to go over two days? Assuming, that is, that you're not crackers?

If we want to see tennis played on a British lawn, perhaps one summer we will collectively fail to be surprised that this will, in 95 per cent of tournaments, mean that rain is part of the event. And people *do* seem to want to see tennis played on a British lawn.

It is as if we have collectively committed to the misapprehension that summers are dry, that we have a rainy season that runs only from September to April. No such thing as bad weather, only the wrong attitude.

As for Glastonbury . . .

We leave London behind, headed still for the place we began. We move more quickly, chasing low pressure. We pass over Berkshire, where a day with thick clouds all around and persistent rain might be called '**cluttery**' and where an alternative to 'rain-drenched' is **zogged** (**#183** and **#184**). Down there is the market town that gave its name to one of the most impressive post-war weather events. In 1959, the Wokingham Storm came: a **high-precipitation supercell** (**#185**), which is like a storm cloud with its own tornado inside and which often looms so large, it makes the sky seem small, unable to contain it.

Below us now, the Hampshire village of Selborne, dotted with devotees of its former resident Gilbert White. A tiny man, White achieved an astonishing amount. He invented birdwatching as we know it, was these islands' first ecologist, is the first person anyone knows of who used the word 'golly' in writing and is credited in the *Oxford English Dictionary* as the first person to use X to end a letter with a kiss.

His meticulous journals are hypnotic in their cycling nature. Here's the start of summer 1776:

> *May 30.* Strawberries blow well. The first effectual rain after a long dry season.
> *June 2.* Sultry, & heavy clouds. Smell of sulphur in the air. Paid for near 20 wasps: several were breeders; but some were workers, hatched perhaps this year.

June 3. Soft rain. Grass & corn improved by the rain already. The long-horned bees bore their holes in the walks.

Not for White the local terms. When the sky is dark and seems certain to rain on Selbourne, he does not record it as **'ter'ble grouty'** (#186); a steady downpour is not described as **'It rained wheelin''** (#187). Just the careful, precise language of a man who notices every effect of every change in the firmament.

Whenever rain stays away, White wishes for its return. And if it should come back with too much enthusiasm for his astonishing garden, he notes that too.

We head south, over the New Forest, where **flisky** (#188) weather is a day covered in mist, but with tiny individual raindrops detectable too. A skittish horse might likewise be flisky.

'There will be rain when Sowley hammer is heard,' they used to remark here: the Sowley ironworks was by the coast and a wind that brought the sound of its industry up through the Forest might presently also bring rain.

AND IN KENT, MORE FOLKESTONE GIRLS THAN EVER

And moving along the south coast now, back in the direction of Devon, a pre-schooler looks out to sea with Grannie and Grandpa.

There is not a cloud up here.

'We'll pay for it later,' offers the pre-schooler, one of those moments where a child echoes some phrase that you don't expect them to use until they are much older. Laughter.

But Grannie and Grandpa feel they should explain a few things.

Grannie remembers 1976, when the Wimbledon grass went brown. Afternoon temperatures averaged 31°C and night after night they dreamed vividly of even a light drizzle the next day.

Grandpa says that, without rain, there'd be no life. Tells tales of how Venus and Mars once had rains, but Venus's boiled away to nothing and Mars's got colder and colder until the last drop fell – then nothing.

Grannie: They *do* say there's no such thing as bad weather, only the wrong clothing. Look at posties. Shorts all year round. Posties know best.

Grandpa: I'll play you a song about rain later by a band we like called the Beatles. It's about how if you try and escape rain, you're not really enjoying life.

Grannie: Because, if you think about it, you love your baths and showers and the water doesn't become nasty just because it's coming from the sky rather than a shower head.

Grandpa: When you live on the British Isles, you know two things: one, that there's going to be plenty of rain; and two, that you don't know when it's going to come.

Grannie: That's right. And so people who complain about rain are really kind of complaining about being alive.

Grandpa: So, what we're saying is: it was funny when you said 'we'll pay for it later', but it is pretty much a silly expression. Where, er, where did you learn it from?

She learned it from Grannie and Grandpa.
More laughter.
Later, when the clouds return and break, all three will hold out their tongues. Salty.

And now we are back approaching Dartmoor.
And there's a mizzle. Actually, or is it a drizzle?

We often hear of bad weather, but in reality, no weather is bad. It is all delightful, though in different ways. Some weather may be bad for farmers or crops, but for man all kinds are good. Sunshine is delicious, rain is refreshing, wind braces us up, snow is exhilarating. As Ruskin says, 'There is really no such thing as bad weather, only different kinds of good weather.'

– from a book about 'how to live', 1894's
The Use of Life by the MP Sir John Lubbock

(The whole thing is almost always misattributed to Ruskin, who himself actually said, in 1883, 'different kinds of pleasant weather'.)

Acknowledgements

Part of this book comes from wet walks with Lucy and Raphael, and rainy days in general with Mum and Alex. I was hugely helped by the staff of the Oxford English Dictionary and the Wimbledon Lawn Tennis Museum, by the Stone Twins, creators of the Irish stamp, by material scientist Veronika Kapsali, by Kirsty McCabe at the Royal Meteorological Society and by a conversation with Chris Marsh about the 'soft day' (all errors are naturally my own). Anna Hervé, David Campbell and Jess Anderson at BBC Books made this real. Thanks to all, and more thanks always to my agent, Andrew Gordon, and even more thanks especially to this book's editor, Shammah Banerjee, who makes things make sense.

It's all dedicated to my late great-aunt Elspeth.

List of Rain Words

1. **mizzle** 4–5
2. **drisk** 6
3. **drizzle** 6–7
4. **kelsher** 7–9
5. **petrichor** 9–10
6. **dimpsy** 11–12
7. **the Brown Willy effect** 13–14
8. **letty weather** 14–15
9. **a fox's wedding** 15–16
10. **sunshower** 15
11. **Ifan y Glaw/John the Rain** 21–2
12. **raining knives and forks** 23
13. **mae hi'n bwrw hen wragedd a ffyn/it's raining old women and sticks** 24
14. **lle diddos** (waterproof place) 25
15. **pistyllio** (rain in the form of a fountain) 25
16. **brasfrwrw** (wide spaced drops) 25
17. **a ffestiniog sgrympian** (a short sharp shower) 25
18. **lluwchlaw** (sheets of rain) 25
19. **chwipio bwrw** (whiplash rain) 25
20. **graupel** 27
21. **virga** 28

22. gwlithlaw 31
23. manlaw 31
24. small rain 32
25. breacbháisteach/half-rain 39
26. pelter 40–1
27. clagarnach 41
28. bull rain 42
29. raining cobblers' knives 42–3
30. raining stair-rods 43
31. soft day 43–4
32. fliuch salach/filthy wet 45
33. fearthainn a rachadh trí chlár darach (rain so persistent that it would go through a board of oak) 45
34. skiffle 45
35. plobán 45
36. crithir 46
37. dry rain 48
38. floggin' 48
39. ag cúr cabáistí/raining cabbages here 48
40. raining upwards 48
41. superhydrophobic 49
42. forlacht 50
43. deluge 50
44. craobhmhúr (a steady quiet sprinkling) 52
45. bucketing 53
46. hammering 53
47. trying to rain 53
48. lashing rain 53
49. fox day 57
50. t'eh ceau fliaghey as cur fliaghey er shen/it's raining vengefully 58
51. claggy 58–9
52. praecipitatio 59

LIST OF RAIN WORDS

53. dree 60
54. comin' deawn full bat 60
55. witchett 61
56. rain shadow 61–2
57. angin 63–4
58. real hunch-weather 66–7
59. just spittin' 68
60. acid rain 70–1
61. peat hags 72
62. grou 78
63. hovering 78
64. church-lead-water 78–9
65. slappy 79
66. washout 79
67. sup o' wet 80
68. yukken it down 80
69. sulphur shower 80
70. grow-rain 81
71. fresh 81
72. water dikes 81
73. damply 81
74. donky day 82
75. parlish slattery 82
76. drap 82–3
77. scotch mist 84
78. sneesl 84
79. groff 86
80. pelsh 86
81. haizer 86
82. probability of precipitation 87–8
83. seepin' 91–2
84. plenit weather 95
85. flobby 96
86. smugly 96
87. bowder 97
88. dreich 97–8
89. thunnerplump 99
90. thunnershock 99
91. blashed 100
92. blirtie 100
93. weetichtie 100
94. skiff 100
95. glashtrochie 101

96.	fairies' baking day 101	117.	timothy 107
97.	plipe 101	118.	thunder-wall 108–9
98.	blurting 101	119.	rain-gös 109
99.	gandiegow 103	120.	shoring 113
100.	greasy 103	121.	weather 113–14
101.	Celtic rainforest 103–4	122.	atween wadders 114
102.	spairging 104	123.	hoyin' it down whole watter 115
103.	sidding 104		
104.	klash 104	124.	stotting 117
105.	roostan hoger 104	125.	dry lightning 118
106.	glaizie 105	126.	plashed 119
107.	murr 105	127.	volutus 120
108.	hagger 105	128.	asperitas 123
109.	daag 105–6	129.	sea fret 124
110.	ug 106	130.	haar 125
111.	slob 106	131.	sylen 125
112.	scouthering 106	132.	evendoon 126
113.	black weet 106	133.	thunder-berries 126
114.	vaandlüb 107		
115.	tümald 107	134.	raggy 128
116.	upslaag 107	135.	luttering 128

LIST OF RAIN WORDS

136. mug 133
137. dringey 134
138. keechy 134
139. mudge 134
140. leaves on the line 134–6
141. heavens-hard 136
142. hollow 136–7
143. black over the back of Bill's mother's 137–8
144. ollin' it dahn 140
145. cow-quaker 140–1
146. plother 141
147. April showers 142
148. arkattit! 143–4
149. ambling 145
150. witter 146
151. wetchered 146
152. spitter 146
153. gobbles 146
154. suent 146
155. tirling 146
156. eevy 147
157. right as rain 149–50
158. bange 154
159. dinge 154
160. lecking time 154
161. bangled 155
162. duggle 158–9
163. peppering 160
164. thight 160
165. sluicy 160
166. mares' tails 164
167. water dogs 164
168. Noe ship 164
169. Sir Roger's blast 164–5
170. rained itself out 166
171. shatter 174
172. geosmin 174
173. blood rain 175
174. to power 177

175. rain 'atchuts an' duckuts 177
176. coming down hasty 178
177. Folkestone girls 178–9
178. ache and pain 179
179. pleasure and pain 179
180. Cynthia Payne 179
181. duke of Spain 179
182. raining cats and dogs 180–2
183. cluttery 186
184. zogged 187
185. high-precipitation supercell 187
186. ter'ble grouty 188
187. rain wheelin' 188
188. flisky 188

About the Author

Alan Connor is an author and screenwriter who has written nonfiction books about the cultures of quizzes and crosswords. His screen credits include *Borat II*, *The Rack Pack* and *A Young Doctor's Notebook* featuring Daniel Radcliffe and Jon Hamm; in 2024 he was nominated for an Emmy for *The Reluctant Traveller*. He has been a writer for *Have I Got News For You* and Charlie Brooker's *Wipe* programmes. Alan is crossword editor for the *Guardian* and has been question editor for *Only Connect* and *House of Games*. He lives and works in Kew, mainly in its Gardens.